Applied Population Ecology

Applied Population Ecology

Principles and Computer Exercises using RAMAS® EcoLab 2.0

Second Edition

H. Reşit Akçakaya
Applied Biomathematics

Mark A. Burgman
University of Melbourne

Lev R. Ginzburg
State University of New York

Sinauer Associates, Inc. • Publishers
Sunderland, Massachusetts

The Cover
Illustrations by Amy Dunham, Department of Ecology and Evolution, State University of New York, Stony Brook. © Applied Biomathematics
Design by Wendy Beck

Applied Population Ecology
   Principles and Computer Exercises using RAMAS® EcoLab
Second Edition

© 1999 by Sinauer Associates, Inc.
All rights reserved.
This book may not be reproduced in whole or in part for any purpose whatever without permission from the publisher.

For information or to order, address:
   Sinauer Associates, Inc.
   P.O. Box 407
   Sunderland, Massachusetts 01375-0407 U.S.A.
   Fax: 413-549-1118, Internet: publish@sinauer.com

RAMAS® EcoLab Software
   © 1990–1995 Electric Power Research Institute
   © 1990–1999 Applied Biomathematics
   RAMAS is a registered trademark of Applied Biomathematics

ISBN 0-87893-028-0

Manufactured in the United States of America

5 4 3 2 1

Contents

Foreword by Robert Goldstein   x
Foreword by Mark Shaffer   xi
Preface   xii
Acknowledgments   xiv

Chapter 1   Population Growth   1
1.1 Introduction   1
    1.1.1 Definition of a Population   2
    1.1.2 Limits to Survival and Reproduction: Niche and Habitat   2
    1.1.3 Mathematical Models in Population Ecology   5
1.2 Births and deaths, immigrants and emigrants   7
    1.2.1 Exponential Growth   8
    1.2.2 Long-lived Species   9
    1.2.3 Using the Model   10
    1.2.4 Doubling Time   12
    1.2.5 Migration, Harvesting, and Translocation   13
1.3 Assumptions of the exponential growth model   14
1.4 Applications   16
    1.4.1 Human Population Growth   16
    1.4.2 Explosions of Pest Densities   21
    1.4.3 Exponential Decline   23
1.5 Additional topic   26
    1.5.1 Population Growth in Continuous Time   26
1.6 Exercises   27
    Exercise 1.1: Blue Whale Recovery   27
    Exercise 1.2: Human Population, 1800–1995   28
    Exercise 1.3: Human Population, 1995–2035   30
1.7 Further reading   31

Chapter 2   Variation   33
2.1 Introduction   33
    2.1.1 Vocabulary for Population Dynamics and Variability   34
    2.1.2 Variation and Uncertainty   36
    2.1.3 Kinds of Uncertainty   36
2.2 Natural variation   37
    2.2.1 Individual Variation   37
    2.2.2 Demographic Stochasticity   38
    2.2.3 Environmental Variation   45
2.3 Parameter and model uncertainty   54
    2.3.1 Parameter Uncertainty   54
    2.3.2 Model Uncertainty   55
    2.3.3 Sensitivity Analysis   56

2.4 Ambiguity and ignorance — 58
2.5 Additional topics — 59
2.5.1 Time to Extinction — 59
2.5.2 Estimating Variation — 60
2.6 Exercises — 61
Exercise 2.1: Accounting For Demographic Stochasticity — 61
Exercise 2.2: Building a Model of Muskox — 62
Exercise 2.3: Constructing Risk Curves — 64
Exercise 2.4: Sensitivity Analysis — 67
2.7 Further reading — 69

Chapter 3 Population Regulation — 71
3.1 Introduction — 71
3.2 Effects of crowding — 72
3.2.1 Increased Mortality — 72
3.2.2 Decreased Reproduction — 73
3.2.3 Self-thinning — 74
3.2.4 Territories — 75
3.3 Types of density dependence — 76
3.3.1 Scramble Competition — 77
3.3.2 Contest Competition — 83
3.3.3 Ceiling Model — 85
3.3.4 Allee Effects — 86
3.3.5 The Concept of Carrying Capacity — 89
3.3.6 Carrying Capacity for the Human Population — 90
3.4 Assumptions of density-dependent models — 91
3.5 Cycles and chaos — 91
3.6 Harvesting and density dependence — 92
3.7 Adding environmental variation — 94
3.8 Additional topics — 95
3.8.1 Equations — 95
3.8.2 Estimating Density Dependence Parameters — 97
3.9 Exercises — 98
Exercise 3.1: Gause's Experiment with *Paramecium* — 98
Exercise 3.2: Adding Stochasticity to Density Dependence — 99
Exercise 3.3: Exploring Differences Between Density Dependence Types — 100
Exercise 3.4: Demonstrating Chaos — 101
Exercise 3.5: Density Dependence and Harvesting — 102
Exercise 3.6: Density *In*dependence Graphs — 104
3.10 Further reading — 104

Chapter 4 Age Structure — 105
4.1 Introduction — 105
4.2 Assumptions of age-structured models — 107
4.3 An age-structured model for the Helmeted Honeyeater — 108
4.3.1 Survival Rates — 109
4.3.2 Fecundities — 111
4.3.3 Sex Ratio — 113
4.4 The Leslie matrix — 113
4.4.1 Leslie Matrix for Helmeted Honeyeaters — 115
4.4.2 Projection with the Leslie Matrix — 117
4.4.3 Stable Age Distribution — 119
4.4.4 Reproductive Value — 121
4.5 Adding Stochasticity — 123
4.5.1 Demographic Stochasticity — 123
4.5.2 Environmental Stochasticity — 125
4.6 Life Tables — 127
4.6.1 The Survivorship Schedule — 128
4.6.2 The Maternity (Fertility) Schedule — 130
4.6.3 Life History Parameters — 131
4.6.4 Life Table Assumptions — 132
4.7 Additional topics — 133
4.7.1 Estimating Survivals and Fecundities — 133
4.7.2 Estimating a Leslie Matrix from a Life Table — 136
4.7.3 Estimating Variation — 141
4.8 Exercises — 143
Exercise 4.1: Building the Helmeted Honeyeater Model — 144
Exercise 4.2: Human Demography — 148
Exercise 4.3: Leslie Matrix for Brook Trout — 149
Exercise 4.4: Fishery Management — 152
4.9 Further reading — 155

Chapter 5 Stage Structure — 157
5.1 Introduction — 157
5.2 Assumptions of stage-structured models — 158
5.3 Stage structure based on size — 159
5.4 A stage model for an Alder — 161
5.5 Building stage-structured models — 163
5.5.1 Residence Times, Stable Distribution, and Reproductive Value — 165
5.5.2 Constraints — 166
5.5.3 Adding Density Dependence — 167
5.6 Sensitivity analysis — 168
5.6.1 Planning Field Research — 168

5.6.2 Evaluating Management Options	170
5.7 Additional topic	171
5.7.1 Estimation of Stage Matrix	171
5.8 Exercises	174
Exercise 5.1: Reverse Transitions	174
Exercise 5.2: Modeling a Perennial Plant	174
Exercise 5.3: Sea Turtle Conservation	176
Exercise 5.4: Sensitivity Analysis	178
5.9 Further reading	181
Chapter 6 Metapopulations and Spatial Structure	**183**
6.1 Introduction	183
6.1.1 Spatial Heterogeneity	185
6.1.2 Habitat Loss and Fragmentation	186
6.1.3 Island Biogeography	187
6.2 Metapopulation dynamics	190
6.2.1 Geographic Configuration	191
6.2.2 Spatial Correlation of Environmental Variation	191
6.2.3 Dispersal Patterns	193
6.2.4 Interaction Between Dispersal and Correlation	196
6.2.5 Assumptions of Metapopulation Models	197
6.3 Applications	199
6.3.1 Reintroduction and Translocation	200
6.3.2 Corridors and Reserve Design	201
6.3.3 Impact Assessment: Fragmentation	202
6.4 Exercises	203
An Overview of the Program	203
Exercise 6.1: Spatial Factors and Extinction Risks	205
Exercise 6.2: Habitat Loss	209
Exercise 6.3: Designing Reserves for the Spotted Owl	210
6.5 Further reading	212
Chapter 7 Population Viability Analysis	**213**
7.1 Introduction	213
7.2 Extinction	214
7.2.1 Extinction in Geological Time	215
7.2.2 Current Extinction Rates	216
7.2.3 The Causes of Extinction	220
7.2.4 Classification of Threat	222
7.3 Components of population viability analysis	224
7.3.1 Identification of the Question and Estimation of Parameters	224
7.3.2 Modeling, Risk Assessment, Sensitivity Analysis	228
7.3.3 Cost-benefit Analysis	228

  7.3.4 Implementation, Monitoring, Evaluation 231
7.4 The limits of population viability analysis 232
7.5 Exercises 234
  Exercise 7.1: Habitat Management for Gnatcatchers 234
  Exercise 7.2: Comparing Management Options 236
  Exercise 7.3: Habitat Loss and Fragmentation 238
7.6 Further reading 240

Chapter 8 Decision-making and Natural Resource Management 241
8.1 Introduction 241
8.2 Detecting impact 242
  8.2.1 Power, Importance, and Significance: An Example 244
  8.2.2 The Precautionary Principle 247
8.3 Managing natural resources 248
  8.3.1 Predicting the Outcome 248
  8.3.2 Explaining the Uncertainty 249
  8.3.3 Model Uncertainty: The Importance of Detail 252
  8.3.4 Strategies and Contingencies 253
8.4 The economic and ecological contexts of natural
   resource management 254
  8.4.1 Uncertainty and Sustainability 256
  8.4.2 The Role of Applied Population Ecologists 257
8.5 Exercises 259
  Exercise 8.1: Statistical Power and Environmental Detection 259
  Exercise 8.2: Sustainable Catch Revisited 261
  Exercise 8.3: Sustainable Use 263
8.6 Further reading 265

Appendix: RAMAS EcoLab Installation and Use 267

References 273

Index 281

Foreword

Protection of endangered or threatened plants and animals is a principal focus in the continuously evolving realm of environmental management policy. Motivation for protecting individual species may arise from a variety of considerations, including esthetic principles, human and ecological health, conservation ideals, biodiversity valuation, and commercial interest. In addition, in order to be effective, environmental policy must balance these concerns with a wider range of technical, economic, and social issues.

An understanding of the basic principles of population ecology is essential to the development of technically sound environmental policy as well as to the creation of specific management strategies for protecting populations. In this text, Akçakaya, Burgman, and Ginzburg—through the use of simple computer simulation models—teach those principles and illustrate their application to a broad spectrum of practical problems. The models used not only take into account the temporal behavior of populations but also the significance of their spatial distribution. A modeling approach is helpful because models are such a powerful means for integrating large amounts of information and data and conducting analyses of uncertainties. Models also provide a means to analyze population responses to alternative management strategies and policies.

A major theme of this text is uncertainty—how to account for it and analyze its implications. For those involved in the development of environmental policy, it is essential to recognize uncertainties inherent in our knowledge of population dynamics, individual species, and the environment. The policy analyst needs to understand the implications of these uncertainties with respect to predicting population behavior. For example, the analyst might need to know whether uncertainties in a specific situation are sufficiently small to permit practical distinctions between predicted results of alternative management scenarios. The analyst also needs to be able to determine what research or monitoring programs would most effectively reduce uncertainty in predictions of a specific population's response.

For over a decade, the Electric Power Research Institute (EPRI) has been funding the development, testing, and application of the RAMAS software used in this book. The motivation for this investment has been to produce risk-based technical tools to address practical questions concerning endangered and threatened species. EPRI supports the publication of this text as a means of transferring the technical knowledge and insights that have been acquired during this process to students, environment professionals, and the general public.

Robert Alan Goldstein
Environment Group
Electric Power Research Institute

Foreword

To many people, population ecology seems, at first acquaintance, to be the antithesis of mathematics. Ecology is about living things, not numbers. It is about the relationships of living things to each other and their environment, not about formulas. Complex things in a complex world that require qualitative observation and description for understanding. Where do deer live? What do they eat? What species of songbirds are found in old-growth forest? Why do some plant species seem to grow everywhere, but others only in specific places? Not the kind of questions that beg for numerical answers.

Eventually, however, as our qualitative knowledge increases, quantitative questions emerge. How many deer can live in a thousand-acre woods? How much food does each need to survive? Why are there more songbird species in larger patches of old-growth forest? How big a patch of old-growth do we need if we want to be sure of keeping all its songbirds around? It does not take long for population ecology to reveal itself as an intensely quantitative discipline.

For whatever reasons, many people drawn to the fascination and beauty of the qualitative aspects of ecology are put off by the quantitative aspects. Mathematics seems far too abstract and inanimate to describe palpable flesh and blood. Yet, it is only through the application of mathematics that we can begin, not just to see, but to understand the underlying patterns in the distribution and abundance of living things that is the essence of population ecology. This book is meant for such people. The text is clear and the examples real. But more than this, the book is accompanied by a friendly computer program that allows the reader to interact with the quantitative aspects of ecology without first having to become a mathematician. A little time with this program and the exercises the authors provide quickly illustrates how dynamic and fascinating quantitative population ecology can be.

Another strength of this text is its scope of coverage. Too many treatments of population ecology start and stop with the basic models of population growth and life tables. This text and computer program capture the basics but go beyond to include such current and difficult topics as metapopulation dynamics and population viability analysis. The authors also provide the best treatment of variation and its effects on population dynamics that I have seen anywhere.

By making this discipline far more accessible to a wider audience, the authors deserve much credit and our sincere thanks.

Mark Shaffer
Vice President for Program
Defenders of Wildlife

Preface

Practical ecological problems such as preservation of threatened species, design of nature reserves, planned harvest of game animals, management of fisheries, and evaluation of human impacts on natural systems are addressed with quantitative tools, such as models. A model is a mathematical representation of a natural process. Many biologists now use models implemented as computer software to approach the quantitative aspects of these practical problems.

In addition to their practical use, such models are excellent tools for developing a deeper understanding of how nature works. You can use the program described in this book, RAMAS EcoLab, to apply most of the concepts discussed in the book and develop your own models. At the end of each chapter, there is a set of exercises. Some of these require only pencil and paper, some require a calculator, and others require the program. Although the book can be used without the program, we believe that most of the more complicated concepts will be much easier to understand when you demonstrate them to yourself using the program.

We hope that, in addition to teaching you the principles of, and practical methods used in, population ecology, this combination of textbook and software will also stimulate you to learn more about modeling, mathematics, and programming. It might even inspire you to write your own computer program for developing ecological models. The principles of building models using a software such as RAMAS EcoLab are the same as those of writing your own equations or computer programs (even though the technical details are very different). Our focus here is not on the mechanics of how a model is implemented, but rather on understanding how various interacting ecological factors should be put together and on understanding the implications of the model's assumptions. Our aim is to discuss principles of population ecology, to show a collection of methods to implement these principles, and to help you appreciate both the advantages and limitations of addressing ecological problems with the help of models.

To the teacher

This book introduces principles of population ecology, with special emphasis on applications in conservation biology and natural resource management. Each chapter includes examples and laboratory exercises based on the software RAMAS EcoLab. While less powerful than the research-grade software developed by Applied Biomathematics, RAMAS EcoLab incorporates all features of the RAMAS Library essential for teaching the basic principles of population ecology, at a level accessible to undergraduate students.

In an introduction to population ecology, most undergraduate students consider learning the mathematics required by traditional texts to be an unnecessary hindrance. The aim of this book is to teach quantitative methods that are necessary to develop a basic understanding and intuition about ecological processes, without intimidating or discouraging students who do not have extensive mathematical backgrounds. Even students who are intimidated by mathematical equations are usually not afraid of using computers. We hope that our integration of software that implements mathematical models in population ecology with an undergraduate textbook will make these models accessible to undergraduates in the biological and environmental sciences.

It should be emphasized that we do not consider developing models with the use of software as an alternative to learning the underlying mathematical concepts. The goal of this book is to introduce mathematical ecology by developing an intuitive understanding of the basic concepts and by motivating the students through examples that put these concepts to practical use. We believe that use of software greatly enhances the understanding of the concepts while encouraging the use of, and emphasizing the need for, quantitative methods.

In addition to the use of software, there are a number of other points in which this text diverges from the more traditional textbooks on population ecology. For example, we decided to develop the models almost exclusively in discrete time (with difference equations) and only briefly mention such things as instantaneous birth and death rates. The equations we use are qualitatively equivalent to the corresponding differential equations, but we believe they are much more intuitive and easy to grasp.

Another important difference is our emphasis on, and early treatment of variability and uncertainty. Use of software instead of analytical models has allowed us to incorporate these important concepts early on, in a way that is simple enough to be easily understood by undergraduate students without strong mathematical backgrounds.

We develop the models from the very beginning in a way that will make the later addition of concepts such as demographic stochasticity and age structure very natural and intuitive. In discussing population regulation, we postponed writing down the famous logistic equation almost to the end of the chapter, concentrating instead on the general, qualitative aspects and dynamic consequences of density-dependent population growth. We started the chapter on age structure with analyzing census data to build a matrix model, rather than the more traditional life-table approach. We suspect that starting with life tables causes some of the confusion that arises when life table variables are to be used to build age-structured models.

We designed the book and the software with sufficient flexibility to allow their use in lecture classes, computer laboratories, or both. They can be used in a lecture class accompanied by a computer laboratory, or in a lecture class in which the examples that require software are assigned as homework exercises, or in a laboratory course where the exercises are the main focus and the conceptual material is read by students. Our hope is that the software tool we provide, in combination with our practical approach, will make population ecology easier to learn and to teach.

Acknowledgments

We thank Martin Drechsler (Botany, University of Melbourne), Alexa Ryhochuk (Zoology, University of Melbourne), Karen Kernan (Applied Biomathematics), and Claire Drill (University of Melbourne) for reviewing the entire book and the software; Soraya Villalba (State University of New York at Stony Brook) for her help in finding many of the examples; Matthew Spencer (Applied Biomathematics) and Sheryl Soucy (SUNY at Stony Brook) for commenting on parts of the book; and Amy Dunham (SUNY at Stony Brook) for the drawings.

Chapter 1
Population Growth

1.1 Introduction

Population ecology is concerned with understanding how populations of plants, animals, and other organisms change over time and from one place to another, and how these populations interact with their environment. This understanding may be used to forecast a population's size or distribution; to estimate the chances that a population will increase or decrease; or to estimate the number of individuals that may be harvested while ensuring a high probability that similar harvests will be available in the future. Thus, the focus of any given study in population ecology may be motivated by very practical considerations in fields as diverse as fisheries harvest regulation, wildlife management, pest control in agricultural landscapes, water quality monitoring, forest harvest planning, disease control strategies in natural populations, or the protection and management of a threatened species. This chapter examines some fundamental concepts in the definition of populations and their environmental limits, and describes first principles of developing population models.

1.1.1 Definition of a Population

The first step in developing an understanding of a population is to define its limits. A population may be defined as a collection of individuals that are sufficiently close geographically that they can find each other and reproduce. The implicit assumption in this definition is that if individuals are close enough, genes will flow among individuals. Thus, it rests on the biological species concept.

In practice, a population usually is any collection of individuals of the same species distributed more or less contiguously. It may refer to a group of individuals in a glass jar, or to a group of individuals that occurs in a conveniently located study area. This approach to the delineation of a populations is particularly pertinent for plant species in which there is vegetative reproduction.

Often, biologists in the field need to determine the geographic boundaries of a population. The limits of a population depend on the size and lifeform of a species, its mode of reproduction, mode of seed or juvenile dispersal, its habitat specificity, and pattern of distribution within its geographic range. Subpopulations may be defined as parts of a population among which gene flow is limited to some degree, but within which it is reasonable to assume that mating is panmictic (i.e., an individual has the same chance of mating with any other individual). All of the factors that make it difficult to define the limits of a population are magnified when trying to determine the limits of subpopulations. Thus, reliance on the underlying principle of reproductive criteria may not be reasonable for some species, and it may not be practical to establish even when sexual reproduction is the dominant mode. In practice, if individuals are grouped and the groups are far enough apart that dispersal or reproduction may be partially limited, we call these groups subpopulations.

1.1.2 Limits to Survival and Reproduction: Niche and Habitat

Animal and plant species are limited in where they can survive and reproduce. Biologists have recognized for many centuries that limits must exist for most species, either in the form of extremes of physical variables or in the form of competitors and predators. The concept of a niche is useful in describing the conditions to which a species is adapted. The niche of a species is its ecological role, its functional relationships with other ecosystem components. It is defined by the limits of ecological variables beyond which the species cannot survive or reproduce. These ecological variables may be abiotic (e.g., temperature, rainfall, concentration of chemicals) or biotic (e.g., food sources, predators, competitors). Each of these variables can be thought of as an axis (Hutchinson 1957). If we focus on two of these variables, the boundaries of the niche may be represented as edges of a rectangle. More usually the edges are drawn as smooth curves. These suggest an interaction between the factors, so that the tolerance to one extreme depends on the levels of the other factors. For example, Figure 1.1 depicts the niche of a species of Sand Shrimp with two ecological variables important for this species: temperature and salinity. Each of these two variables is represented by one axis of the graph. The lines represent percentage mortality, which is lowest when both salinity and temperature are moderate. Extremes of both salinity or temperature cause increased mortality. The response of a species to one niche variable will depend to some degree on the values of the other variables. For example, the Sand Shrimp might tolerate a higher temperature if salinity is within the optimal range (Figure 1.1).

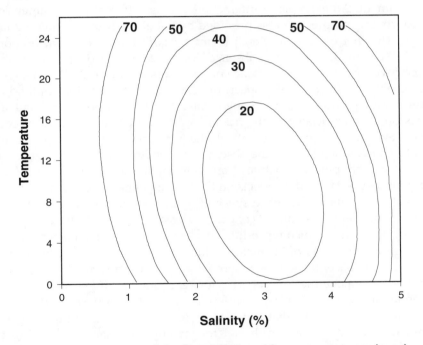

Figure 1.1. The niche of the Sand Shrimp (*Crangon septemspinosa*) in terms of temperature (°C) and salinity (%), under conditions of low concentrations of dissolved oxygen. The curves represent percentage mortality. After Haefner (1970).

The *fundamental niche* is the niche defined by all the abiotic environmental variables that affect a species. It represents the limits of physical conditions that a species can tolerate. The *realized niche* is defined by both biotic and abiotic variables. It includes things such as food availability, tolerable physical conditions, competition with other species for resources (such as nest sites or nutrients), and the avoidance of predators. The niche may be determined by the availability of potentially limiting resources, or by physical properties or disturbances that limit a population. For example, the quality of a particular part of the niche space may depend on the abundance of predators and their ability to exploit that niche. The niche of a species may vary in time and space because the physiological or behavioral properties of individuals in the population may differ at different times and in different places.

The size and shape of the niche will change through time, responding to changes in the properties of individuals in the population, as well as in the environment. An example of interaction among different niche dimensions (food, cover, and predators) is provided by the effect of predation on young perch. The diet of juvenile Perch (*Perca fluviatilis*) changes from predominantly copepods in the absence of predators to predominantly macroinvertebrates in the presence of predators (Figure 1.2). This change in the niche preferences of the species is caused by the fact that individuals forgo foraging opportunities in open water when predators are present, and focus on prey associated with structures offering protection such as rocks and crevices. Such dynamics affect the interactions among species that would compete for food in the absence of predators. Thus, structural complexity of the physical habitat can determine the composition of fish communities because of its effects on the feeding behavior of young fish.

Not all points in the niche space of a species are equally conducive to survival and reproduction. In concept, at least, the space includes a preferendum, a region in which reproduction and survival are maximized. Beyond this region, the quality of the niche declines monotonically to the boundaries of the niche space, to regions where survival and reproduction are barely possible. The niche that is necessary for regeneration and survival through juvenile life stages is usually different and frequently somewhat more restricted than the niche that is necessary for survival as an adult. The environmental limits that an adult can tolerate may be narrower during the reproductive season than during the rest of the year.

The habitat of a species is the place where the species lives. It is a geographical concept, the place in which the set of conditions necessary to support a species exists. The environmental and biotic variables that define the niche of a species are not fixed, but change in time and space. Thus, a place that was habitat in one year may not be habitat in the next if the requisite environmental conditions no longer exist at that place; the individuals that live there may move, fail to reproduce, or die. Drought, for example,

Introduction 5

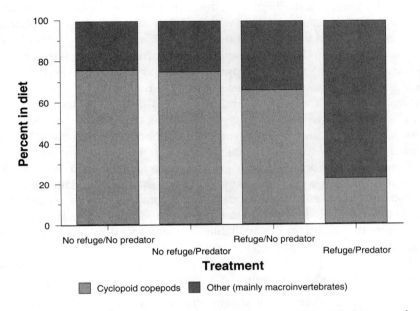

Figure 1.2. The diet of juvenile Perch in the presence and absence of competition, and the presence and absence of a refuge (after Persson and Eklov 1995). The refuge was a structurally diverse substrate and offered a number of feeding opportunities that were not present in open water. The results in the figure were obtained from replicated laboratory experiments.

will contract the geographic range in which a species can survive and reproduce. In such circumstances, the niche of the species is not altered, but the habitat contracts. In this book, we examine population ecology with a particular focus on understanding and explaining a species' response to changes in its environment, including its habitat.

1.1.3 Mathematical Models in Population Ecology

Population ecology, as we discussed above, is concerned with changes in the abundance of organisms over time and over space. Abundance and how it changes can be described by words such as "abundant" or "rare," and "fast" or "slow," but population ecology is fundamentally a quantitative science. To make population ecology useful in practice, we need to use quantitative methods that allow us to forecast a population's future and express the results numerically.

Frequently, the need to make forecasts leads to the development of models. A model is a mathematical description of the population. A model may be as simple as an equation with just one variable, or as complex as a computer algorithm with thousands of lines. One of the more difficult decisions in building models (and one of the most frequent mistakes) concerns the complexity of the model appropriate for a given situation, i.e., how much detail about the ecology of the species to add to the model.

Simple models are easier to understand and more likely to give insights that are applicable in a wide range of situations. They also have more simplistic assumptions and lack realism when applied to specific cases. Usually, they cannot be used to make reliable forecasts in practical situations.

Including more details makes a model more realistic and easier to apply to specific cases. However in most practical cases, available data are limited and permit only the simplest models. More complex models require more data to make reliable forecasts. Attempts to include more details than can be justified by the quality of the available data may result in decreased predictive power and understanding.

The question of the appropriate level of complexity (i.e., the trade-off between realism and functionality) depends on:

(1) characteristics of the species under study (e.g., its ecology),

(2) what we know of the species (the availability of data), and

(3) what we want to know or predict (the questions addressed).

Even when detailed data are available, general questions require simpler models than more specific ones. For example, models intended to generalize the effect of one factor (such as variation in growth rate) on a population's future may include less detail than those intended to forecast the long-term persistence of a specific species, which in turn, may include less detail than those intended to predict next year's distribution of breeding pairs within a local population.

The purpose of writing a model is to abstract our knowledge of the dynamics of a population. It serves to enhance our understanding of a problem, to state our assumptions explicitly, and to identify what data are missing and what data are most important. If the data required for building the model are plentiful, and if our understanding of the dynamics of a population are sound, we may use the model to make forecasts of a population's size or behavior. In the rest of this chapter, we will introduce some very simple models. In later chapters, we will add more details to these models.

1.2 Births and deaths, immigrants and emigrants

The Muskox (*Ovibos moschatus*) is a large mammal that was eliminated from substantial areas of its natural range in North America and Greenland during the 1700s and 1800s by excessive hunting. The last individuals on the Arctic Slope of Alaska were killed in 1850–1860. In 1930, the legislature of the Territory of Alaska authorized funds to obtain stock from Greenland for reintroduction to Nunivak Island. The island was to serve as a wildlife refuge in which the reintroduced population could grow. It was chosen because it was relatively accessible, was free of predators, and permitted confinement of the population to a large area of apparently good habitat. A population of 31 animals was reintroduced to Nunivak Island in 1936. Once grown, the population was to serve as a source for further reintroductions on the Alaskan mainland.

The population was censused irregularly between 1936 and 1947, and then annually between 1947 and 1968 (Figure 1.3). The objective of this section is to use the census information to construct a model that may be useful in managing the population.

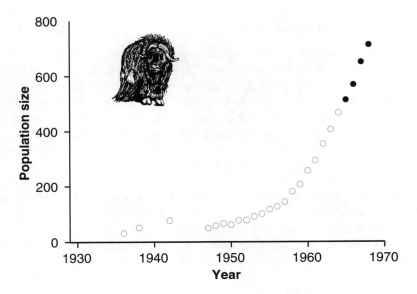

Figure 1.3. Population censuses of Muskox on Nunivak Island between 1936 and 1968. During the last four years of the census, from 1965 to 1968 (shown as closed circles), some animals were removed from the island and relocated to new sites (after Spencer and Lensink 1970).

The total number of individuals (N) in a fixed region of space can only change because of births, deaths, immigration and emigration. Change in population size over a discrete interval of time from t to $t+1$ can be described by the equation

$$N(t+1) = N(t) + B - D + I - E$$

where B and D are the total number of births and deaths respectively during the time interval from t to $t+1$, while I and E are the total number of individuals entering and leaving the region during the same time interval (the time interval in this example is one year). Of course, we may replace immigration and emigration by processes that are mediated by humans, such as reintroductions, harvesting, or poaching. Change in population size from year t to $t+1$ is given by $N(t+1) - N(t)$.

Many populations like the Nunivak Island Muskox population are closed in the sense that there is no immigration or emigration. In these cases the model for the population becomes

$$N(t+1) = N(t) + B - D$$

1.2.1 Exponential Growth

Rather than express births and deaths as numbers of individuals, we may express them as rates. For an annual species, the formulation of an equation to express population growth is relatively simple. In an annual species, all the adults alive at year t die before year $t+1$. Thus the number of individuals in the population next year is equal to the number in the population this year, multiplied by the average number of offspring per individual,

$$N(t+1) = N(t)\,f$$

We express births as the fecundity rate, f. It may be thought of as the average number of individuals born per individual alive at time t that survive to be counted at the next time step, $t+1$.

For an annual species, fecundity is equal to the growth rate of the population,

$$N(t+1) = N(t)\,R$$

The rate of population increase (or, population growth) is conventionally represented by the symbol R. It is called the "finite rate of increase" of a population. Despite the use of words such as "growth" and "increase" in its definition, R can describe both growth and decline in abundance. If R is greater than 1, the population will increase, and if R is less than 1, the population will decrease. When births and deaths balance each other, R equals 1, and the population abundance stays the same.

If we want to predict the population size for two years, instead of one, we can use the above equation twice:

$$N(t+2) = N(t+1)\ R$$
$$N(t+1) = N(t)\ R$$

If we combine these two equations,

$$N(t+2) = N(t)\ R\ R$$
$$N(t+2) = N(t)\ R^2$$

More generally, if there is a need to predict the population size t time steps into the future beginning from time step 0, the equation for population growth may be written as

$$N(t) = N(0)\ R^t$$

which says that, to estimate the population size at time step t, multiply the population size at the beginning, $N(0)$, with the growth rate, R, raised to the power of t ("raised to the power of t" means multiplied by itself t times).

This equation represents a model for the dynamics of a population. A simple model such as this one is an equation describing the relationship between independent variables, parameters, and dependent variables. A dependent variable (or state variable) is the quantity you want to estimate (such as the future population size). It depends on the other factors, called independent variables. Parameters are those components of a model that mediate the relationship between independent and dependent variables. The equation above allows us to estimate the population size at any time in the future. The population size at time t, $N(t)$, is the dependent variable, and time (t) is an independent variable. The growth rate (R) and the initial population size, $N(0)$, are parameters. The type of population growth described by this model is called *exponential growth* because of the exponentiation in R^t. Sometimes, it is also called *geometric growth* or *Malthusian growth* (after Thomas Malthus).

1.2.2 Long-lived Species

Many species are not annual; they survive for more than one year and reproduce more than once. To allow for the survival of individuals for more than a single time step, we may introduce a survival rate (s), which is the proportion of individuals alive at time t that survive to time $t+1$. Thus the population size at the next time step is the sum of two numbers: (1) the number of individuals that survive to the next time step (out of those that were already in the population), and (2) the number of offspring produced by them that survive to the next time step. We can write this sentence as the following formula

$$N(t+1) = N(t)\,s + N(t)\,f$$

This is a model of population growth in which births and deaths are expressed as fecundities and survivals. By rearranging the formula we get

$$N(t+1) = N(t)\,(s + f)$$

The sum $s + f$ represents the combined effect of fecundities and survivals on the population abundance. If we add these two numbers, we can rewrite the equation as

$$N(t+1) = N(t)\,R$$

where R is the same growth rate that we discussed above. In the rest of this chapter, we will only use R and not concentrate on its components s and f. These components will become important in the next chapter, because the effect of certain types of variability depends on how the growth rate is partitioned into survival and reproduction. We will further develop this distinction between survival and reproduction when we learn about age-structured models in Chapter 4.

1.2.3 Using the Model

The first task in applying the model above is to estimate R. We may rearrange the equation so that

$$N(t+1)/N(t) = R$$

Let's consider the growth of the Muskox population on Nunivak Island (Figure 1.3). Knowing the sizes of the Muskox population in 1947 and 1948 (49 and 57, respectively), we can estimate the growth rate of the population in 1947 simply as 57/49, which equals 1.163. This estimate of R does not use all of the information available to us. We know the population sizes in all years between 1947 and 1968. Figure 1.4 below provides values for R for all years during which observations were made in consecutive years.

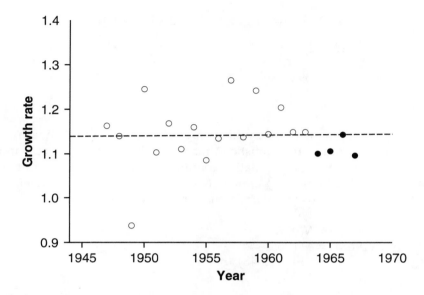

Figure 1.4. Growth rates of the Muskox population on Nunivak Island between 1947 and 1968. The closed circles represent years in which animals were removed from the population. These rates were excluded from the estimation of the average growth rate for the population, represented by the dashed line.

Perhaps the most striking aspect of Figure 1.4 is that the growth rate of the population is not fixed, but varies from one year to the next. There is no apparent trend through time. We will explore the causes and consequences of this variation in the next chapter. For the moment, if we want to make forecasts about the size of the population, it will be necessary to calculate the average growth rate of the population. The growth of a population is a multiplicative process. That is, we estimate next year's population by multiplying the current population by the average growth rate. Because growth is a multiplicative process, the appropriate way to calculate the average growth rate is to find the geometric mean of the observed growth rates. There are 17 growth rates between 1947 and 1964, so we calculate the geometric mean of the series by multiplying these 17 numbers, and taking the 17th root of the result:

$$R^{17} = (1.16 \times 1.14 \times 0.94 \times 1.25 \times 1.1 \times 1.17 \times 1.11 \times 1.16 \times 1.09 \times 1.14$$
$$\times 1.27 \times 1.14 \times 1.24 \times 1.14 \times 1.2 \times 1.15 \times 1.15)$$
$$= 10.392$$
$$R = \sqrt[17]{10.392}$$
$$= 1.148$$

The population increased by an average of 14.8% per year between 1947 and 1964. We can use this statistic to make predictions for the population. For example, if the island population continues to grow at the same rate as it has in the past, what is the population size likely to be in 1968, given that there were 514 Muskox in 1965? To answer this, we would calculate

$$N(t) = N(0) \, R^t$$
$$N(1968) = N(1965) \cdot R^3$$
$$= 514 \cdot (1.148)^3$$
$$= 777.7$$

Thus, in three years, we can expect that the population will increase to about 778 animals.

If a population is growing exponentially, then the population size should appear to be linear when it is expressed on a log scale. The log scale is a means for verifying visually that a population is indeed growing exponentially. The Muskox data fit a straight line quite well when population size is on a log scale (Figure 1.5).

1.2.4 Doubling Time

Frequently, the rate of increase is expressed in terms of doubling time. Given that the average rate of increase is known, how long will it take for the population to double in size? We know from above that

$$N(t) = N(0) \cdot R^t$$

where $N(0)$ is the current population size and $N(t)$ is the population size t time steps in the future. We may write the question above as: If $N(t)/N(0)$ equals 2, what is t? In other words, if

$$R^t = 2$$

then what is t? Taking the natural logarithms of both sides of the equation, and rearranging, we get

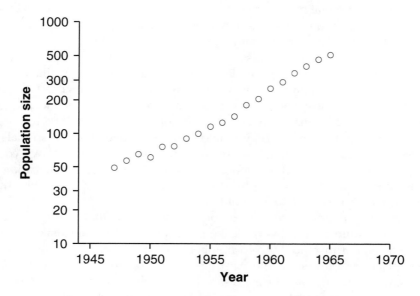

Figure 1.5. Plot of population size versus time for the Nunivak Island Muskox population. Note that the population size (y-axis) is in logarithmic scale.

$$t \cdot \ln(R) = \ln(2)$$
$$t = \ln(2) / \ln(R)$$

In the case of the Muskox, the doubling time is $\ln(2)/\ln(1.148)$, which is equal to 5 years.

1.2.5 Migration, Harvesting, and Translocation

We estimated above that the population in 1968 would be close to 778 individuals. In fact, the population size was recorded at 714 animals. Note that the growth rate of the population in the period from 1965 to 1968 was consistently below average (Figure 1.4). At least part of the reason for this was that 48 animals were removed from the population and released in sites elsewhere in Alaska, a process called translocation. Without these removals, the population would have been 762, much closer to the predicted value of 778. Of course, this ignores mortality and reproduction among those animals that were removed, had they remained.

One may use calculations of the growth rate of the population to estimate the number that may be removed in perpetuity. The basic idea is that if things continue in the future as they have in the past, one could remove or harvest a number of animals from a population so that the effective growth rate is 1. In the case of the Muskox, the natural rate of increase is 1.148, so 0.148 (or about 15%) of the population, on average, will be available. In a population of 714 animals, this would be about 106 animals. We have already seen that the growth rate varies from year to year, which means that the number of additional individuals in the population may not be 106 every year. A more intelligent strategy would seek to remove all those animals in excess of a base population of, say, 700 animals. Nevertheless, we could expect to have about 100 animals available each year for translocation to other sites. If we were managing the population for harvest rather than translocation, the calculations would be the same.

If the population were not completely closed, so that immigration or emigration could occur, this could be incorporated into the existing model quite easily. For instance, a fixed number may arrive on average each year, but a fixed proportion of the existing population may disperse to other places. Immigration could be expressed as the addition of a fixed number (I), because the number that arrives in a population often does not depend on what is already in the population. Emigration could be expressed as a rate (e), if we assume that more animals would emigrate from a more crowded population. The model would become

$$N(t+1) = N(t)\,s + N(t)\,f - N(t)\,e + I$$
$$= N(t)\,(s + f - e) + I$$

In this model, the emigration rate is incorporated into the growth rate term, R. Emigrants are treated in the same way as deaths.

In some cases, it may be better to represent emigration as a fixed number rather than a rate. For example, the management strategy for a species may stipulate a fixed number of removals. However the model is written, it should best reflect the dynamics of the species in question. There are no strict rights and wrongs in building mathematical expressions to represent the dynamics of a population. The only rule is that they should represent the real dynamics of the population as faithfully as possible.

1.3 Assumptions of the exponential growth model

Whenever a model is constructed, it employs a set of assumptions reducing the complexity of the real world to manageable proportions. Assumptions are all those things not dealt with explicitly in the model but that must nevertheless be true for the model to provide reasonably accurate predictions. The model above makes a number of assumptions; it is obviously a vast oversimplification. A list of assumptions of this model is:

(1) There is no variability in model parameters due to the vagaries of the environment. The model for exponential population growth above is clearly a deterministic model; there is no uncertainty in its prediction. It says that at some time, t, in the future, the population size will be $N(t)$, and it can be calculated exactly by the right-hand side of the expression. We have already seen that the rates are not likely to be constant over time. We will explore the consequences of varying growth rates in the next chapter.

(2) Population abundance can be described by a real number. In other words, the model ignores that populations are composed of discrete numbers of individuals. In fact, birth and death rates (and immigration and emigration rates) may vary simply because real populations are discrete and structured. We will explore this factor in the next chapter. However, this kind of variation is unimportant in large populations.

(3) Populations grow or decline exponentially for an indefinite period. This implies that population density remains low enough for there to be no competition among members for limiting resources. These processes (related to density-dependent effects) are discussed in Chapter 3.

(4) Births and deaths are independent of the ages or of any other unique properties of the individuals. Essentially, we assume that individuals are identical. In real populations, the probability of surviving, the number of surviving offspring produced, and the propensity to immigrate or emigrate are not likely to be the same for different individuals in a population. They may depend to some extent on the age, sex, size, health, social status, or genetic properties of the individuals. However, it turns out that even if birth and death rates are, say, age-dependent, the mean rate per individual will remain constant if the proportions of the population in each age class remain constant over time (see Chapter 4). Thus, it is sufficient to assume that the proportion of individuals in these different categories (such as age) remains the same. This assumption will be violated by, for example, genetic changes in the population, or by changes in the sex ratio (the relative numbers of males and females), or by changes in the age structure (the relative numbers of individuals at different ages). In Chapters 4 and 5, we will discuss models that track changes in the composition of a population.

(5) The species exists as a single, panmictic population. Within the population, the individuals are mixed. The interaction with other populations of the same species is characterized by the rates of emigration (as a constant proportion of the abundance per time step) and immigration (as a constant number of individuals per time step). The interactions with other species are characterized by their constant effects on the population growth rate. In Chapter 6, we will explore models in which the dynamics of several populations of the same species are simultaneously described.

(6) The processes of birth and death in the population can be approximated by pulses of reproduction and mortality; in other words, they happen in discrete time steps and are independent. Of course, the time interval may be made arbitrarily short, in which case the models would approach formulations in continuous time. The continuous time analogues of these expressions will be explored in Section 1.5.

1.4 Applications

We have already explored some applications of the exponential growth model in the above discussion. These applications are relevant to wildlife management, translocations and reintroductions, and harvesting control. In this section we describe applications to human population projections and pest control.

1.4.1 Human Population Growth

The exponential model for population growth is so simple that one might hesitate to use it in any real circumstances. However, we have seen that it approximates the population dynamics of the Muskox population reasonably well, at least in the short term. It was invented originally by Malthus in 1798 to predict the size of the human population. It still fits the growth of human populations, both globally and within individual countries. Table 1.1 shows estimates of the human population size in the recent past.

In the 45 years between 1950 and 1995, the population grew from 2.51 billion to 5.75 billion ("billion" has different meanings in different countries; here we use 1 billion = 10^9 = 1,000 million). Using these figures, and the equation

$$N(t) = N(0) \cdot R^t$$

or

$$N(1995) = N(1950) \cdot R^{45}$$

we can calculate the annual rate of growth. Rearranging the equation,

Table 1.1. Estimates of the human population size.

Year	Population in billions (i.e., $\times 10^9$)
1800	0.91
1850	1.13
1870	1.30
1890	1.49
1910	1.70
1930	2.02
1950	2.51
1970	3.62
1975	3.97
1980	4.41
1985	4.84
1990	5.29
1995	5.75

After Holdren (1991), and Pulliam and Haddad (1994).

$R^{45} = N(1995) / N(1950)$
$R^{45} = 5.75 / 2.51 = 2.29084$,

which means that R, multiplied 45 times by itself equals 2.29084. To find the value of R, we need to find the 45th root of 2.29084, or find

$R = 2.29084^{(1/45)}$

You can do this with a calculator or a computer (using spreadsheet software, for example). Another easy way is to use logarithms. Taking the logarithm of both sides, simplifying, and then exponentiating, we get

$\ln(R) = (1/45) \cdot \ln(2.29084)$
$\ln(R) = 0.02222 \cdot 0.82892$
$\ln(R) = 0.01842$
$R = \exp(0.01842) = 1.01859$

These figures suggest a rate of increase per year between 1950 and 1995 of about 1.0186, or 1.86% per year. At this rate, the human population doubles every 37.6 years. The rate between 1800 and 1950 was about 1.0068, or 0.68% each year (which corresponds to a doubling time of 102.5 years). The increase in the growth rate between these two periods is generally attributed to improvements in medicine and standards of living that have served to reduce the death rate.

Predicting the size of the human population holds a great deal of interest for organizations that deal with public health, international trade, and development planning. The models that are used for these purposes are more complex than the ones developed so far in this book, but they use the same kinds of parameters such as fecundities, survivorships, and migration. They differ mainly by assuming that growth rates are not fixed but will change over the coming few decades, and by assuming that the growth rate will decline to 1.0 in most countries in the first half of the next century.

A few projections of the human population have been carried out without assuming that the human rate of increase will slow down, resulting in clearly unreasonable predictions. For example, if the growth rate of 1.0186 was to be sustained until the year 2100, there would be 40 billion people on earth, and by 2200 there would be 262 billion people. There is little doubt that the planet cannot sustain this many people, even given the most benign assumptions about the interactions between people and the environment. Using a different model in which the rate of population increase was itself increasing, Von Foerster et al. in 1960 suggested that the human population would become infinite in size on November 13, 2026 (this date is the so-called Doomsday prediction). The United Nations takes a more conservative view, assuming appropriate slowing in the growth rate of human populations, and it predicts a total population size of between 7.5 and 14.2 billion people by the year 2100. This view rests on the assumption that the birth rate in many nations will decline to equal the death rate over the next few decades (Figure 1.6). The reasoning is that there is a certain amount of cultural inertia that results in large family sizes, developed originally to compensate for child mortality, and that this propensity towards large families will erode over one or two generations as people realize that most of those born will survive.

The size of the human population is also of considerable interest to ecologists and wildlife managers, not least because of the relationship between the size of the human population and the rate of the use of natural resources, both within most countries and globally. For example, collection of firewood and charcoal for domestic and industrial use is an important cause of forest clearance, particularly in savanna woodlands, and increasingly in tropical moist forests. However, the relationship is not simple. Frequently, the unequal distribution of land and other resources plays an important part in determining the rate at which resources are used.

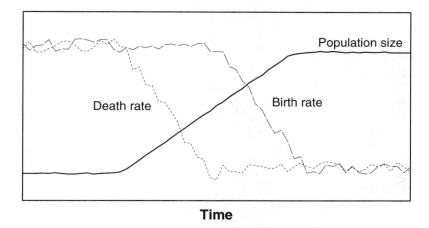

Figure 1.6. The "benign demographic transition" assumed in models used to predict human population growth in the 21st century (after Hardin 1993). The difference between the death and birth rates causes population growth, but birth rates eventually decline and the population growth stops.

Human population growth raises three issues. They include the absolute size of the population (P), the per capita consumption of biological resources (called affluence, A), and the environmental damage of the technologies employed in supplying each unit of consumption (T). The impact (I) of the human population on the natural environment may be expressed as (see Ehrlich and Ehrlich 1990; Hardin 1993)

$$I = P \times A \times T$$

Human population size and energy use are relatively easy to measure (Table 1.2, Figure 1.7). Environmental impact per unit of consumption is more difficult. Energy use per person has risen over the last 150 years. Even if the environmental impact per unit of consumption remains constant, changes in environmental impact are measured by the product of increasing affluence and increasing population size.

Of course, the equation above is a vast oversimplification of the issue of human population growth. For instance, it ignores interactions and cumulative effects that may be felt long after the impact is made. Humans make direct or indirect use of about 30% of the terrestrial net primary production of the planet, and the changes caused by human impacts have reduced ter-

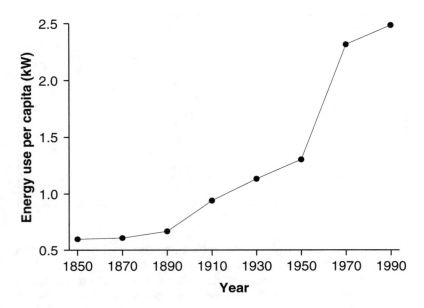

Figure 1.7. Energy use per person, 1850–1990 (after Holdren 1991 and Vitousek 1994).

restrial net primary production by about 13%. As a species, humans currently usurp approximately two-fifths of the productivity of terrestrial ecosystems. The product of population size, per-capita consumption and environmental damage per unit of consumption sets the limit to human activities. It remains to be seen exactly what that limit is. Per-capita consumption of energy is one indicator of environmental load, and recent estimates for different parts of the world are provided in Table 1.2. Japan is considered to be an industrial country that is an efficient energy user. The United States alone has been responsible for 30% of the world's cumulative use of industrial energy forms since 1850.

Malthus pointed out about 200 years ago that increasing human populations would eventually create unsupportable demands on natural resources. Even in countries with little or no population growth, per-capita consumption grows more or less exponentially. The solutions to many problems lie in managing resource consumption and in the equitable distribution of access to and use of resources. In the case of nonrenewable resources, net consumption eventually will have to be reduced to zero. It is relatively easy to become pessimistic about the way in which humans use natural resources,

Table 1.2. Population size and average energy consumption for different demographic groups in 1990.

Population	Per-capita energy consumption (gigajoules per year)	Population size (millions)
Industrial countries	205	1210
Developing countries	22	4081
India	11	855
Africa	25	550
Japan	134	150
Australia	230	18
North America	330	280
Global average	63	

After Holdren (1991), and Boyden and Dovers (1992).

although there are still many people who believe in the ultimately benign nature of human development and interactions with the environment. Despite such optimism, it is likely that as the human population and its standard of living increase, the effects of human activities on the earth's resources will accelerate in the near future.

1.4.2 Explosions of Pest Densities

Ecological explosions are rapid, large-scale and frequently spectacular increases in the numbers of a species. They have had long-term and important impacts on ecosystems and on the health and economic well-being of human communities. The term explosion was coined in reference to plant and animal invasions by Charles Elton in 1958, to describe the release of a population from controls. Diseases often show explosive growth. Influenza broke out in Europe at the end of the First World War and rolled around the world. It is reputed to have killed 100 million people. The rabbit viral disease myxomatosis is a nonlethal disease of Cotton-tail Rabbits in Brazil, which has a lethal effect on European rabbits. It was introduced to Europe in the early 1900s and eliminated a great part of the rabbit population of Western Europe. It was also introduced as a biological control agent of feral rabbit populations in Australia, dramatically reducing rabbit populations there after its introduction in the early 1950s. Explosions also occur when plants and animals are deliberately or accidentally introduced to islands and continents where they did not exist before. There were several attempts to introduce the Starling (*Sturnus vulgaris*) to the United States from Europe in the 1800s. A few individuals from a stock of 80 birds in Central Park, New York, began breeding in 1891. By 1910, the species had spread from New Jersey and Connecticut. By 1954, the species spread throughout North America as far as the West Coast of the United States and northern Canada, and was beginning to invade Mexico.

Agricultural systems and natural communities in many countries are threatened by the introduction of pests and diseases that, in their country of origin, are harmless or tolerable annoyances. We live in a period of the world's history when the rate of movement of species among continents and between regions is perhaps higher than it has ever been, largely as a result of deliberate translocations by people.

Exotic animal populations usually are defined as pests because they damage production systems such as crops or livestock. Similarly, a weed is simply any unwanted plant, usually defined by its impact on productive systems. Environmental pests and weeds are species that invade natural communities, changing the composition or adversely affecting the survival of the native biota. Pests and weeds may be the result of an introduction from another region or country, or they may be local (endemic) species that have become more abundant because of changes in the landscape or because of natural cycles in the population. Most deliberate introductions between continents or regions have been for ornamental or utility reasons. Movement within continents may involve natural dispersal (by wind or water), animal movements (native, domestic, and feral animals disperse the seeds of weeds), vehicles, transport of agricultural products and so on.

Frequently, explosions in the population size of pest animals take the form of an exponential increase. For example, in 1916 about a dozen individuals of the Japanese Beetle (*Popillia japonica*) were noticed in a plant nursery in New Jersey. In the first year, the beetle had spread across an area of less than a hectare. By 1925, it had spread to over 5,000 km^2, and by 1941 it had spread to over 50,000 km^2 (Elton 1958).

The beetles probably arrived in 1911, on a consignment of ornamental plants from Japan. In Japan, they were seldom a pest, held in check by their own natural predators, competitors, diseases, and limited resources. In America, their numbers became formidable. By 1919, a single person could collect 20,000 individuals in one day. The species fed on and often defoliated over 250 species of plants, including native North American plants, and many commercially important species such as soy beans, clover, apples, and peaches.

Genetic improvement of species, by selection and field trials, has long been a focus of agricultural science. The development by plant molecular biologists of transgenic crop plants that are resistant to predation by insects and infection by fungal and viral pathogens is an area of active research, and one with immediate potential for economic gain. Targets for research include the development of plants for pesticide resistance, nitrogen fixation, salt tolerance, and tolerance to low nutrient status. Many results of transferring genes between species may be environmentally beneficial. For example, development of innate pest resistance will decrease the dependence of agriculture on pesticides. Genetic engineering also provides advantageous prospects for wildlife management, such as fertility control of feral animals and the biological control of weeds.

However, genetically engineered organisms pose risks through (1) the effects of transgenic products (primary and secondary); (2) the establishment and spread of transgenic crop plants in nontarget areas; and (3) the transfer, by hybridization and introgression, of transgenes from crops to wild, related material. It is typically difficult or impossible to predict the effect of the products of a transgenic species on the multitude of species and processes with which the species will come into contact. Most predictions for the likelihood of transgenic plants forming feral populations assume that their potential is the same as other exotic species and that, if successful, such species are likely to spread in an exponential fashion, at least in the short term. Such dynamics may be adequately modeled by the exponential growth model.

1.4.3 Exponential Decline

It is well known that human exploitation of marine mammal populations has resulted in steady declines. The harvesting pressure on many species continued over much of this century until the animals became so scarce that it was no longer economically viable to catch them. Several important whale fisheries have followed this pattern, including Fin, Sei, and Blue Whales, all baleen whales of the Antarctic Ocean. We will explore the dynamics of the Blue Whale population, using available data to fit a model of exponential decline.

For Antarctic whales, virtually complete and reasonably accurate data of the catches are available (Figure 1.8). A decline in Blue Whale stocks was clearly evident from catch data before the Second World War. The war resulted in a cessation of whale harvest, which commenced again in earnest after 1945.

The declines in stocks had been a cause for concern among whaling nations. The International Whaling Commission, set up in 1946, set limits to the total Antarctic catch. The Blue Whale catch was largely replaced by Fin Whale catches after 1945, as Blue Whales became rarer. The general quota provided no differential protection of species, and there was provision for revision of the quota if there were declines in stocks.

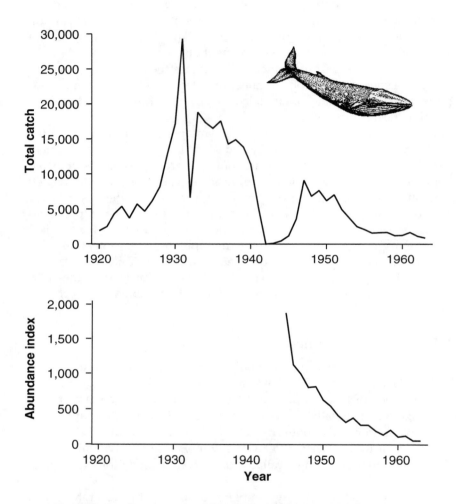

Figure 1.8. Total catch of Blue Whales in the Antarctic, 1920–1963, and an index of abundance of Blue Whales in the Antarctic (estimated as the number of whales caught per catcher-ton-day), 1945–1963 (after Gulland 1971).

Blue Whale populations were already very depleted by the time quotas were introduced in 1945. The stocks of the species continued to decline, and a shorter open season for the species was introduced in 1953. However, the difference between the catch and the productive potential of the whale population continued to widen because the quotas were more or less fixed and the population did not reproduce quickly enough to replace the numbers removed. The imposition of quotas, and the allocation of the catch among countries, were topics of intense political and scientific argument from the 1950s through 1967. In 1960 and 1961, the International Whaling Commission failed to set quotas at all because of disagreements among its member nations. As late as 1955 there was no agreement on the extent, or even the existence, of a decline in Antarctic whale stocks. Fin Whales were the mainstay of the industry at this time, and their abundance did not begin to decline dramatically until 1955, even though the abundances of other whale species were obviously falling. Throughout the period of the early 1960s, Blue Whale stocks continued to decline. The population abundance data for the Blue Whale from the period 1945–1963 fit a straight line quite well, suggesting that the decline in the population size was approximately exponential (Figure 1.9).

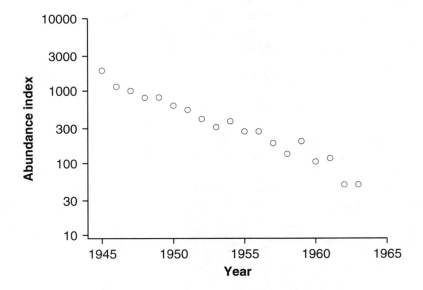

Figure 1.9. The abundance index for Blue Whale in Figure 1.8 plotted on a logarithmic scale.

In 1963, evidence was presented to the whaling industry that its quota was three times higher than the level at which further depletion of the stock could be avoided. The industry reduced its harvest to these levels by 1967. One critical failure in the process of regulation of the industry was that scientists failed to provide clear advice to the industry after 1955, when a reduction in the quota was clearly necessary and would have been much less drastic than the reduction that eventually was necessary. The Blue Whale population was reduced from about 20 to 50 thousand individuals in the 1930s to between 9 and 14 thousand in the mid 1950s. It remained approximately constant at about 14,000 individuals between 1965 and 1975.

1.5 Additional topic

1.5.1 Population Growth in Continuous Time

Most examples in this book involve populations of species living in temperate regions, which have distinct reproductive seasons tied to the seasonality of the environment. This property, together with the way most field studies estimate demographic parameters (by periodically observing a population), make it easy and natural to use the discrete-time formulations of population models. However, some natural populations reproduce and die continuously, as does the human population. The basic model for population growth in discrete time was

$$N(t+1) = N(t) + B - D$$

This could be rewritten as

$$\begin{aligned}\Delta N &= N(t+1) - N(t) \\ &= B - D\end{aligned}$$

The symbol ΔN is the difference in population size. If the time interval represented by ΔN is small, we can approximate it by the derivative dN/dt. Rather than express births and deaths as numbers of individuals, they may be expressed as instantaneous rates, giving

$$\begin{aligned}dN/dt &= bN - dN \\ &= (b - d)N \\ &= rN\end{aligned}$$

The difference between the birth rate and the death rate in continuous time is called the instantaneous growth rate (r). The equation above may be solved, giving

$$N(t) = N(0) \cdot e^{rt}$$

This equation says that the population size at time t in the future is given by the current population size, multiplied by e^{rt}. In this equation, e is a constant (about 2.7); sometimes e^{rt} is written as $\exp(r \cdot t)$. By analogy with the equivalent discrete time equation, you can see that

$$R = e^r$$

because $R^t = (e^r)^t = e^{rt}$. The equation for exponential population growth in continuous time is equivalent to the model in discrete time, in which the time interval is made arbitrarily small. Frequently, models for population growth are written in continuous time because they are analytically tractable, i.e., one can find solutions to the equations using calculus. Equations in discrete time, although more plausible for many biological scenarios, are generally less tractable. However, this is not a big disadvantage when numerical solutions can be obtained using computer simulations. We will ignore models in continuous time in this book because discrete-time models are more applicable to most of our examples, and they are easier to explain and understand. While we shall mention analytical solutions where they exist, we will use computer simulations to solve most of the problems.

1.6 Exercises

Exercise 1.1: Blue Whale Recovery

This exercise is based on the Blue Whale example of Section 1.4.3. The population dynamics of the Blue Whale population and predictions of harvest levels have been made using exponential models. The growth rate (R) of the population during the period represented in Figure 1.9 was 0.82, i.e., the population declined by 18% per year. The fecundity of Blue Whale has been estimated to be between 0.06 to 0.14 and natural mortality to be around 0.04. In the absence of harvest, the growth rate of the population would be between 1.02 and 1.10. We want to estimate the time it will take for the Blue Whale population to recover its 1930s level. Assuming a population size in 1963 of 10,000 and a target population size of 50,000, calculate how many years it will take the population to recover:

    (a) if its growth rate is 1.10
    (b) if its growth rate is 1.02

Hint: Use the method for calculating doubling time, but with a factor different from 2.

Exercise 1.2: Human Population, 1800–1995

In this exercise, we will investigate the data on human population growth given in Section 1.4.1. Before you begin the exercise, look at your watch and record the time.

Step 1. Calculate the growth rate of the human population for each interval in Table 1.1. Note that each interval is a different number of years: initially 50, then 20, later 5 years. It is important to convert all these into annual growth rates, so that we can compare them. Use the method described in Section 1.4.1 to calculate the annual growth rate from 1800 to 1850, from 1850 to 1870, so on, and finally from 1990 to 1995. Enter the results in Table 1.3 below (in the table, the first growth rate is already calculated as an example).

Table 1.3. Calculating the annual growth rate of the human population.

Year	Population (billions)	Time interval (years)	Population in previous census	Growth rate in T years (R^T)	Annual growth rate (R)
t	$N(t)$	T	$N(t-T)$	$N(t)/N(t-T)$	$[N(t)/N(t-T)]^{(1/T)}$
1800	0.91	—	—	—	—
1850	1.13	50	0.91	1.24176	1.00434
1870	1.30		1.13		
1890	1.49		1.30		
1910	1.70		1.49		
1930	2.02		1.70		
1950	2.51		2.02		
1970	3.62		2.51		
1975	3.97		3.62		
1980	4.41		3.97		
1985	4.84		4.41		
1990	5.29		4.84		
1995	5.75		5.29		

Step 2. Plot the growth rate against year and comment on any pattern.

Step 3. It is important to know the difference between relative and absolute growth. Even though the annual growth rate (a relative measure of growth) declines, the number of individuals added to the population each year (an absolute measure of growth) may increase. The number added to the population in one year is equal to $N \cdot (R-1)$, where N is the population size and R is the annual growth rate. For example, in 1850,

$$1.13 \text{ billion} \cdot 0.00434 = 4.9 \text{ million}$$

people were added to the population. (Strictly speaking this is not correct, because the two numbers refer to different times: 1.00434 is the average growth from 1800 to 1850, whereas 1.13 billion is the population size in 1850. However, for the purpose of this exercise, it is a reasonable approximation.)

Calculate the number of people added to the human population each year, for 1975, 1985, and 1995, using Table 1.4 below. Compare the change in annual growth rate with the absolute increase in the population size per year.

Table 1.4. Calculating the number of individuals added to the human population.

Year	Population size	Annual growth rate	Annual number of people added to the population
1975	3.97 billion		
1985	4.84 billion		
1995	5.75 billion		

Step 4. Using the estimated number of people added to the human population in 1995, calculate the approximate number of people added to the human population:
  (a) per day
  (b) per hour
  (c) per minute
  (d) during the time you completed the exercise

Exercise 1.3: Human Population, 1995–2035

In this exercise you will investigate one rather optimistic scenario of the slow-down and stabilization of human population. Specifically, you will calculate the population size in 2035, assuming that by that time the growth rate has reached 1.00 (no growth). For this exercise, assume that (i) the fecundity in 1995 is 0.0273, (ii) the survival rate will not change in the future, and (iii) in the 40 years after 1995, the fecundity will decrease so as to make the annual growth rate in 2035, $R(2035) = 1.0$.

Step 1. Using the annual growth rate for 1995 you calculated above, estimate the annual decrease in fecundity necessary to make $R(2035) = 1.0$. Assume a linear decrease, i.e., an equal amount of decrease in fecundity for each year.

Step 2. Calculate the fecundity and the annual growth rate for years 2005, 2015, 2025, and 2035, and enter them in Table 1.5 below.

Step 3. Calculate the 10-year growth rates for the periods 1995–2005, 2005–2015, 2015–2025, and 2025–2035, by multiplying each annual growth rate by itself 10 times. For example the 10-year growth rate for 1995–2005 is $R(1995)^{10}$. Enter these in the table below (enter the 10-year growth rate for period 1995–2005 in the line for 1995.)

Table 1.5. Projecting human population growth.

Year	Fecundity (f)	Annual growth rate (R)	10-year growth rate (R^{10})	Population at the beginning of the 10-year interval	Population at the end of the 10-year interval
1995	0.0273			5.75 billion	
2005					
2015					
2025					
2035		1.0000	1.0000		

Step 4. Estimate the population size at the end of each 10-year period by multiplying the 10-year growth rate you calculated in the previous step with the population size at the beginning of the time period.

How much did the population increase while the fecundity was decreasing for 40 years? If the fecundity decreased to the same level in 80 years instead of 40, would the final population size be larger or smaller?

1.7 Further reading

Ehrlich, P. R. and A. H. Ehrlich. 1990. *The population explosion.* Simon and Schuster, New York.

Elton, C. S. 1958. *The ecology of invasions by animals and plants.* Methuen, London.

Hardin, G. 1993. *Living within limits: ecology, economics and population taboos.* Oxford University Press, New York.

Holdren, J.P. 1991. Population and the energy problem. *Population and Environment* 12:231-255.

Vitousek, P. M., P. R. Ehrlich, A. H. Ehrlich and P. M. Matson. 1986. Human appropriation of the products of photosynthesis. *BioScience* 36:368–373.

Chapter 2
Variation

2.1 Introduction

We view population ecology as an applied science that helps find solutions to practical problems in wildlife and game management, natural resource management and conservation, and other areas. All of the cases explored in Chapter 1 dealt with real world problems. Yet they ignored a fundamental component of the ecology of populations, namely variability in populations and in the environment in which they live. Such variation is pervasive. The growth rate of the Muskox population on Nunivak Island varied substantially around the average of 1.148 that was used to predict future population sizes. The rate of decline in the Blue Whale population averaged 0.82 between 1947 and 1963, but in no year was it exactly 0.82. In this chapter, we introduce the concepts and the framework that are necessary to deal with natural variation in population ecology.

Ecologists think in terms of what is known as the central tendency of their data. The first questions to come to mind in any population study usually are of the kind: "What is the average growth rate?" A somewhat more thoughtful ecologist might also ask "What is the year-to-year variation in the growth rate?" or even "What are the confidence limits on the predicted population size?" These are all important concepts. It is equally important to consider the distribution of outliers. In practical situations, for example, it is often important to know the worst case we might expect, and how likely it is. The chances of extreme events are particularly relevant to people interested in keeping population sizes within predetermined limits. To look at data or to make predictions in this way first requires a special vocabulary.

2.1.1 Vocabulary for Population Dynamics and Variability

Stochasticity is unpredictable variation. If the long-term growth rate is less than 1.0, the population will become extinct, no matter how stable the environment. These populations are said to be the victims of "systematic pressure"; their decline results from deterministic causes. Populations that would persist indefinitely in a constant environment nevertheless face some risk of extinction through variation in fecundity and survival rates. These populations, when they decline, are the victims of stochasticity.

In Chapter 1, we began constructing models to represent the dynamics and ecology of populations. Population models that assume all parameters to be constant are called deterministic models; those that include variation in parameters are called stochastic models. Stochastic models allow us to evaluate the models in terms of probabilities, accounting for the inherent unpredictability of biological systems. The probabilities generated by stochastic models allow us to pose different kinds of questions. We might want to know the worst possible outcome for the population: If things go as badly as possible, what will the population size be? We might like to know which parameter is most important. When the problems that we face are subject to uncertainty (and they almost always are), then the questions we ask should be phrased in a specific way. For example, if our focus is the size of the population, then we should ask:

> What is the probability of {decline / increase} to [population size, N_c] {at least once before / at} [time, t] ?

The components inside braces {...} are choices and the components inside square brackets [...] are quantities. Circumstances will ordain whether we are most interested in (or concerned about) population increase or population decline, or both. We must specify the critical population size or threshold (N_c) that represents an acceptable (or unacceptable) outcome, or a range of such values. We must specify a time horizon (t), a period over which we wish to make predictions. Lastly, we must say whether it is sufficient that these conditions are met at least once during the period or that they are met at the end of the period.

The words risk and chance may be used in place of the word probability, but they emphasize slightly different aspects of a problem. Risk is the potential, or probability, of an adverse event. When applied to natural populations of plants and animals, risk assessment usually is concerned with the calculation of the chance that threatened populations will fall below some specified size, or that pests will exceed some upper population size. Declines in population size may be seen as desirable when dealing with a pest, in which case we talk of reduction. They may be undesirable when dealing with rare species, in which case we may refer to the risk of decline or risk of extinction. Similarly, increases may be either desirable (recovery of rare or threatened species) or undesirable (explosion of pest species). If we wish to estimate the chance of decline or increase of a population to some specified size (a threshold) at least once in the specified period, we talk of the "interval" probability. If our interest is in the chance of being above or below a threshold at the end of the time horizon, we talk of the "terminal" probability.

The critical population size, or threshold, specified in the definition of risk often reflects an abundance that is considered to be too low (for rare or threatened species) or too high (for pest species). It may be determined on an economic basis for harvested species, for example, when a fishery manager wants to maintain a certain population of Brook Trout in a stream.

Over a given time period, there is a chance that any population will become extinct. This chance we term the background risk. If the consequences of different types of human impact are measured in terms of probabilities, it is possible to compare them against the background risks that a population faces in the absence of any impact. Added risk is the increase in risk of decline that results from some impact on a natural population. Similarly, if the consequences of different types of conservation measures are measured in terms of probabilities, we can compare them against the background risks in the absence of conservation efforts. The difference (which we hope is a decrease in the risk of decline) is a measure of the effectiveness of the conservation effort.

The probability of extinction or explosion of a population in a given time period is one way we can describe the chances faced by natural populations. Another way is to use the expected time to extinction or explosion. These statistics are the average time it takes for a population to fall below or to exceed specified population thresholds. We will discuss this further under the heading Additional topics (Section 2.5.1).

2.1.2 Variation and Uncertainty

We saw in Chapter 1 that the change in size of a population is governed by births, deaths, immigrants, and emigrants. Births and deaths may be governed by environmental parameters. Variation in the environment leads to variation in survival and fecundity rates, and results in variation in population size that is independent of the average growth rate of the population. "Good" years are those in which the population produces more offspring and experiences fewer deaths. Species respond to environmental variation in different ways. The time scales of impact and response are related to the ecology of populations. Some species will resist environmental change and others will respond to it, depending on its severity and duration.

The picture is further complicated by the fact that estimates of population size will vary from one time to the next, even in the absence of any real change, because of measurement errors. Further, some populations will fluctuate in a regular fashion, following diurnal, seasonal, or longer term weather patterns, or because of their interactions with predators or competitors. Natural variation in the environment and measurement error will overlay any other natural or human caused patterns, trends, or cycles in population size. The consequences of this variation are that we cannot be certain what the population size will be in the future. In addition, there are other factors that may cause population sizes to vary unpredictably, and there are other reasons why our predictions may be uncertain. However, if we can characterize this uncertainty, we may be able to provide an indication of the reliability of any estimate that we make. We will explore these concepts below and introduce ways of dealing with them in circumstances where predictions are necessary for resource and wildlife management and species conservation.

2.1.3 Kinds of Uncertainty

Uncertainty may be considered to be the absence of information, which may or may not be obtainable. Uncertainty encompasses a multiplicity of concepts including:

incomplete information (what will the population size be in 50 years?)

disagreement between information sources (what was the population size last year?)

linguistic imprecision (what is meant by the statement "the population is threatened"?)

natural variation (what will be the depth of snowfall this winter?)

relationships between variables (does resistance to cold in winter depend on the amount of food available in the preceding summer?)

the structure of a model (should emigration be represented as a number or a rate?)

Models are simplifications of reality. Uncertainty may be about the degree of simplification that is necessary to make the model workable and understandable. It may be about the decision we should take, even if all other components of the problem are known or understood. Different types and sources of uncertainty need to be treated in different ways. Probability may be a useful means of describing some kinds of uncertainty. Others are more appropriately handled with decision theory, or even with political process. There are numerous classifications of the kinds of uncertainty and variability. Decomposing uncertainty into its different forms allows us to use available information together with appropriate tools to make predictions. These predictions may then be qualified by a degree of uncertainty.

2.2 Natural variation

2.2.1 Individual Variation

Individual variation is the variation between individuals within the same population. It is the term used to describe the variation within a population due to genetic and developmental differences among individuals that results in differences in phenotype. Individual variation also includes genetic variability. Each individual has a different genetic makeup that results from the combinations of genes in its parents, and the random selection of those genes during meiosis. The rate of change in the genetic makeup of a population is inversely proportional to the number of adults that contribute to reproduction. In small populations, the genetic composition of the population may change significantly because of these random changes, a process known as genetic drift.

Inbreeding is mating between close relatives. In small populations, mating between relatives becomes more frequent. If the parents are related to one another, rare recessive genes are more likely to be expressed and genetic variation may be lost. These processes may alter the survival and fecundity of individuals, and reduce the average values of these rates in the population as a whole. The loss of variation may also reduce the ability of the population to adapt to novel or extreme environmental conditions.

Other differences among individuals contribute to this kind of uncertainty. For example, in species with separate sexes, uneven sex ratios may arise by chance and have an enhancing or a detrimental effect on further population increase. While these processes are relatively well understood, it is not possible to say if, and to what extent, these effects will be felt in any given instance. The process is inherently unpredictable.

2.2.2 Demographic Stochasticity

Demographic stochasticity is the variation in the average chances of survivorship and reproduction that occurs because a population is made up of a finite, integer number of individuals, each with different characteristics. Consider the following example. The Muskox population on Nunivak Island began in 1936 with 31 individuals and had an average growth rate of 1.148. On the basis of this average, we might expect the population in 1937 to be 35.6, but there is no such thing as 0.6 of a Muskox. The growth rate we specified is an average based on observations. What this result says is that, 4 to 5 more births than deaths are expected in the Muskox population between the 1936 census and the 1937 census. Exactly how many, we cannot be sure.

The people who conducted the censuses of the Muskox population on Nunivak Island recorded the number of calves produced each year. Over the years the average number of calves per individual (f) was 0.227. Given that

$$R = f + s$$

the average survival rate was

$$s = 1.148 - 0.227 = 0.921$$

The parameters in the models we developed in Chapter 1 do not vary, so they are termed deterministic models. They provide a single estimate of population size at some time in the future. We could add an element of realism to these models by following the fate of each individual. For example, rather than multiplying the whole population by a survival value of 0.921 to calculate the number of survivors, we could decide, at each time step, whether each individual survives or dies. We do this in such a way that, in the long term, 92.1% of the individuals survive. One way to do this is to choose a uniform random number between 0 and 1 for each individual. ("Uniform" means that each number in that range has an equal chance of being sampled; see the exercises section for ways of choosing random numbers.) If the random number is greater than the survival rate of 0.921, then the individual dies. Otherwise, the individual lives. We ask the question for each individual in the population, using a different random number for each. Thus, if there are 31 individuals in the population, there is no guarantee that 29 will survive, although it is the most likely outcome ($31 \times 0.921 = 28.6$). There is some smaller chance that 28 or 30 will survive and some still smaller chance that 27 or 31 will survive. This kind of uncertainty represents the chance events in the births and deaths of a real population, and is what we mean when we talk of demographic uncertainty.

We could add a further element of reality by treating the births in the population in an analogous fashion. Like deaths, births come in integers (no Muskox will produce 0.227 offspring: rather, most will produce none, some 1). We can represent this in our model by following the fate of individuals in the same way as we did for survival. That is, choose a random number for each individual. If the value is less than 0.227, the animal has an offspring. Otherwise, it does not.

A time step of a year seems appropriate because reproduction in this species is seasonal and the environment is highly seasonal. We treat the population as composed of an integer number of individuals and we sample the survival and reproduction of members of the population, using the observed population size and the population average fecundity and survival rates. The result is that our predictions will no longer be exact. As in a real population, our model reflects how a run of bad luck could lead to the extinction of any population, no matter how large the population size or how large the potential growth rate.

Each time we tally the population and we ask "Does this animal die?" and "Does this animal produce offspring?", the answer may be different. To gain some idea of the expected outcome, and the reliability of that outcome, we need to run a series of trials. We need to repeat the experiment a number of times and calculate the average and the variability of the outcome. Variability of a set of numbers is often expressed as their variance or standard deviation (variance is equal to the standard deviation squared). Histograms showing the frequencies of different possible population sizes one year after the introduction of Muskox to Nunivak Island are shown in Figure 2.1. The larger the number of trials, the more reliable will be our knowledge of the average and the variance. This approach is most effectively implemented on a computer.

Formulating demographic stochasticity in this way makes a number of assumptions about the ecology of the population. It assumes that a female can have no more than one offspring per year. More efficient and more general methods are available that involve sampling the binomial and Poisson distributions, but learning how to use them is beyond the scope of this book

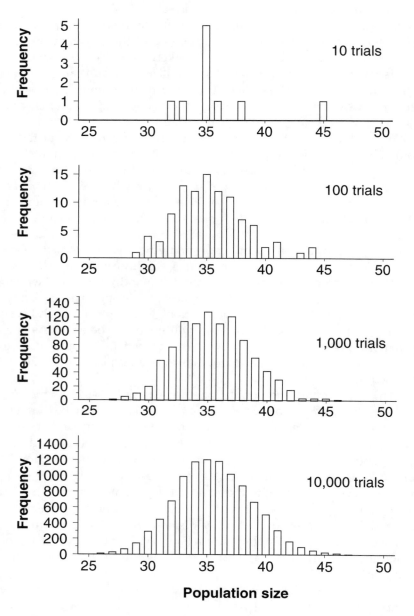

Figure 2.1. Histogram of population sizes for a Muskox population model with demographic stochasticity.

(the computer program that comes with this book implements these more advanced methods). Our approach here assumes further that births and deaths are independent events. We choose different random numbers to represent the survival and reproductive success of each individual. If an animal dies in 1937, it may also have offspring before it dies that year. We could, if we wanted, preclude reproduction if an animal dies, or make it less likely for an animal to survive if it reproduces.

It is clear that demographic stochasticity can have an important effect on estimates of population size. From a starting population of 31, the population quite reasonably could increase to 46 animals, or decrease to 27 animals after one year, just because of the random chances associated with giving birth and surviving from one year to the next. This kind of variability is present in every population. The deterministic expectation of 35.6 is just one of many possible outcomes. The mean predicted by the model including demographic stochasticity (Figure 2.1) is similar to the number predicted by the deterministic model (35.6). By carrying out a great many trials, we can be reasonably certain that we know the mean and the variation in expected population sizes. The uncertainty arises because real populations are structured, composed of discrete individuals, and because the individual occurrences of births and deaths are unpredictable.

By developing forecasts in this way, we can ask different kinds of questions. For example, we could ask "How likely is it that there would be less than 31 animals in 1937?" or "What is the chance that the population will increase by 30% or more, rather than the average 14.8%?" To answer these questions let's count the number of trials that met the stated criteria and divide by the total number of trials. For example, to answer the first question, we tally the number of trials that reached 30 animals, 29 animals, etc., down to the smallest recorded number (which was 24). The result is given in the second column of Table 2.1. The third column shows the cumulative frequencies, i.e., the number of trials predicting a given number *or fewer* individuals. Each row of this column is calculated by adding up the numbers in the second column up to and including the current row. Adding up the first 7 numbers gives 548, which is the number of trials in which the predicted number of animals was 30 or less. The last column gives the same (cumulative number), divided by the total number of trials (10,000 in this example). Note that this table contains only part of the data represented by the last histogram in Figure 2.1; the dots ("..") at the end of table are to remind you that the maximum population size was 49, and the table could have another 18 rows.

Table 2.1. Number of trials (out of the total of 10,000 trials) that predicted 24 to 31 animals in 1937.

Population level (N_c)	Number of trials that reached a level $\leq N_c$	Cumulative number of trials (that reached a level $\leq N_c$)	Probability of decline to N_c
24	3	3	0.0003
25	2	5	0.0005
26	11	16	0.0016
27	29	45	0.0045
28	67	112	0.0112
29	142	254	0.0254
30	294	548	0.0548
31	449	997	0.0997
..	..	..	..

According to the table, 548 trials out of 10,000 predicted a population size of 30 or less, so the chance is 548/10000 or 0.0548. Thus, even though the deterministic model tells us the population will increase, and the stochastic model tells us the population will probably increase, there is a better than 5% chance that the population will actually decline from 1936 to 1937.

We can answer the second question posed above in a similar way. The question was "What is the chance that the population will increase by 30% or more?" An increase of 30% is equal to a population size of 40.3. The number of trials that predicted a population size greater than 40 was 669. The chance of exceeding 40 is therefore 0.0669, or about 6.7%. Note that you cannot find this answer in the table above. The above table shows the probability of reaching a level *less than or equal to* N_c, whereas this question was expressed in terms of reaching a level *greater than or equal to* N_c.

The task of wildlife managers is to implement plans to manage both the expected population size and the probabilities of extreme outcomes. Wildlife management questions that may be answered by population forecasts come basically in two forms. The first is: "What is the chance that the population will exceed some threshold?" (for control problems) and the second is "What is the chance that the population will fall below some threshold?" (for conservation problems). The management of natural populations may require ensuring that the populations remain within prespecified levels, so that both the upper and the lower bounds are important. For example, large herbivores in parks or reserves frequently must be maintained within upper and lower limits so that they persist indefinitely within the confines of the reserves without becoming so numerous that they displace other herbivores.

Alternatively, it may be important to manage various ecological processes and human impacts to maintain a population, to keep it from becoming extinct.

To address these questions, we may redraw the histograms in Figure 2.1 as cumulative frequencies. As we demonstrated above, if the cumulative frequencies are divided by the number of trials, they may be interpreted as probabilities. Thus, the curve in Figure 2.2 represents the chances that the population which began as 31 individuals in 1936, will be equal to or less than various threshold population sizes in 1937. The x-axis of this curve is the threshold population size (first column of Table 2.1), and the y-axis is the probability that the population size will be less than or equal to the threshold (last column of Table 2.1).

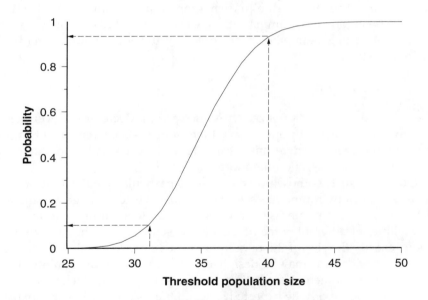

Figure 2.2. Cumulative frequencies from Figure 2.1, divided by 10,000 (the number of trials), and plotted against population size.

To interpret the figure above, let's use the two sets of arrows on the figure to answer a couple of questions: What is the chance that the population will be equal to or less than 31 individuals in 1937 (in other words, what is the chance of no increase)? Looking at the figure, we see that the curve predicts a probability of about 0.1, or 10% for a threshold population size of 31 (see also the last column of Table 2.1, which shows a probability of 0.0997, or about 10%, of declining to 31 or below). What is the chance that the population will be less than or equal to 40 individuals in 1937? The answer is about 0.93. The curve represented in Figure 2.2 is called a risk curve. More specifically, it is a quasi-extinction risk curve. It provides answers to questions phrased as follows: "What is the chance that a population with current size N will fall below some critical threshold population size, N_c, within the next period, t?" Thus, it is useful for questions concerning the lower bound of population size.

Demographic stochasticity, as well as phenotypic variation of all kinds, has most important consequences in small populations. This is because the effects are inversely related to population size. We can see the qualitative effect of population size by considering the survival probability for the Muskox, 0.921. Assume some catastrophe affects the population and only two animals remain. What is the chance that both will die before the following year? The chance due to demographic uncertainty is $(1 - 0.921)^2 = 0.0062$. When there are 31 animals, the chance is $(1 - 0.921)^{31}$, which is a very small number. In general, the chance of loss of the entire population (p) in a single time step is

$$p = (1 - s)^N$$

where N is the population size. As N increases, p decreases. Nevertheless, even for medium-sized populations, there remains some chance of important deviation from the deterministic model and some small chance of loss of the population through nothing more than bad luck.

Questions such as those posed above are particularly relevant to wildlife managers and environmental scientists who have to manage populations within limits. They are phrased and answered quite naturally in terms of the probabilities of different outcomes. Common sense tells us that we can never predict exactly the size of the population next year. Models that include elements of randomness may be designed to cope with the uncertainty that is part of all environmental prediction and decision making. Such models will allow us to target both the expected size and the risk of decline or expansion of a population. We will see below that, to some extent, these properties are independent. The management strategies to maximize the expected population size may be different than those that are required to minimize the risk of decline.

It is important to remember that, even though the models we developed in this section allowed variability in the number of survivors or the number of offspring, they did not allow the survival rates and fecundities to vary. We demonstrated that *even when these rates remain the same*, demographic stochasticity introduces randomness and unpredictability in the estimated population size. In the next section, we will add more realism to our models by allowing their parameters (survival rates and fecundities) to vary.

2.2.3 Environmental Variation

2.2.3.1 Temporal variation

Environmental variation is unpredictable change in the environment in time and space. It is most often thought of as temporal variation at a single location. An obvious example is rainfall. Even in circumstances in which we know precisely the average annual rainfall of a location based on records going back centuries, it is difficult to say if next year will be relatively wet or dry, and even if next week will be rainy or not.

In circumstances in which the vital rates of a population depend on environmental variables, the rates will likewise be unpredictable. The concept of a niche implies that a set of biotic and abiotic variables limits the distribution of a species. It is usually assumed that a set of environmental parameters will affect the rate of growth of a population within the niche that a species occupies. Environmental variation that results in variations in population size is seen as a mechanism that is extrinsic to the population. Environmental variation is not the sole determinant of fluctuations in population size. We will explore intrinsic causes of population change in subsequent chapters.

Environmental variation results in fluctuations in population size when environmental variables affect the number of survivors and the number of offspring in a population. There are many examples of relationships between environmental variables, and the survival and fecundity of individuals within populations. For example, population numbers of the California Quail are influenced by climate. High winter and spring rainfall is associated with high reproduction in semi-arid regions (Figure 2.3). The mechanisms for this dependence may be based on the quality and quantity of plant growth or the availability of free drinking water. If water is scarce in the region inhabited by the California Quail, fewer juveniles survive than if water is plentiful.

The causes of interactions between population dynamics and environmental variables such as rainfall may be less direct than in the example above. The fecundity of Florida Scrub Jays, expressed as nest success, is likewise dependent on rainfall (Figure 2.4). However, the researchers speculate that the direct cause of variation in nest success is variation in nest predation rates. Rainfall could influence nest predation by affecting the density or activity of predators, the availability of alternative food items, the nest vigilance of the Jays, or the protective vegetation cover surrounding nests.

46   Chapter 2 Variation

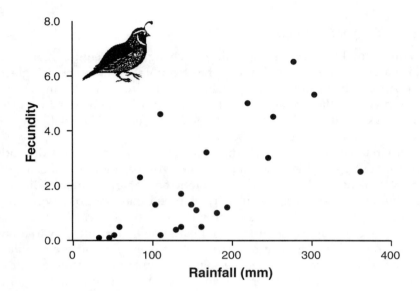

Figure 2.3. The relationship between rainfall (December to April precipitation) and fecundity in the California Quail (*Callipepla californica*) for a population in the Panoche Management Area, California (after Botsford et al. 1988). The correlation coefficient for these (log transformed) data was 0.68. Fecundity was expressed as the number of juvenile birds per adult.

There are many causes of death in the Muskox population on Nunivak Island, some of which are directly related to environmental variables. Over the 20-year period that observations were made, animals fell from cliffs, became lost on sea ice, were mired in a bog, drowned, were otherwise injured, were shot by humans, or died during winter snow falls. There were almost certainly deaths due to starvation in years of heavy snowfall, during which it was harder to find food. A relatively common event in this population was for small groups of animals to wander onto pack ice around the island during winter. The ice floes broke up or melted, blocking the animals' return to land. These animals either starved or were drowned at sea. It would be impossible to predict the number of animals that might suffer such a fate in any year, because it depends on the propensity of groups to wander over the ice, and the chance environmental events that lead to the break up

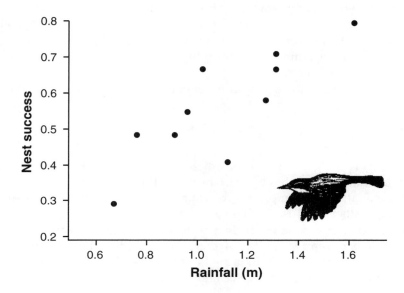

Figure 2.4. Nest success in Florida Scrub Jays (*Aphelocoma c. coerulescens*) as a function of total rainfall in the preceding 10 months (June to March) (after Woolfenden and Fitzpatrick 1984). Nest success is the proportion of nests that survive to fledgling. The correlation between rainfall and nest success is 0.78.

of the pack ice. Weather conditions are thought to be the single most important factor determining year to year variation in population growth of Muskox on other islands (see Gunn et al. 1991).

If we wanted to predict the population size next year, and in making this prediction take into account the variation due to some environmental factor, we would need to know three things: (1) which environmental factor is important, (2) how it affects the population dynamics, and (3) what the value of that environmental factor will be in the future. In other words, even if the dynamics of a population are directly related to an environmental variable (and we knew exactly what this relationship is), we still cannot make precise predictions because it is impossible to say what the value of the environmental variable will be next year.

We noted in Chapter 1 that the growth rate of the Muskox population was not fixed through the period of observation. It varied from a maximum of 1.27 to a minimum of 0.94. Having taken note of the fact, we estimated the mean growth rate and then made some predictions for population sizes that ignored the fact that growth rates are variable. The results of our predictions were made without any estimate of how reliable they were. For example, we predicted that the population size in 1968 should have been 778 animals. It turned out to be 714 (or 762 if you include removed animals). Was the prediction within the bounds of probability, given the variable nature of the population's growth rate?

We may rewrite the equation for exponential population growth as follows:

$$N(t+1) = N(t) \cdot R(t)$$

where $R(t)$ is the growth rate for time step t. Writing $R(t)$ instead of R indicates that the growth rate varies from one time step to the next. When we use this equation, we sample the growth rate from some distribution for each time step, rather than use a fixed value. We may, for example, use observed distribution of growth rates for the population (Figure 2.5). This distribution shows that between 1947 and 1964, there was one year when the growth rate was between 0.90 and 0.95 (indicated at the mid-value of this range, 0.925), one year when it was between 1.00 and 1.05, etc.

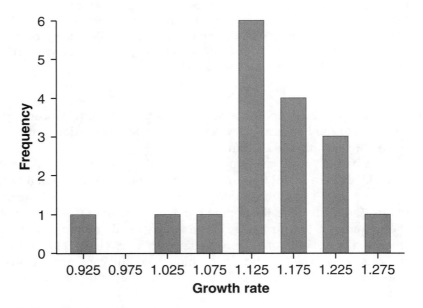

Figure 2.5. Frequency distribution of growth rates observed in the Muskox population on Nunivak Island between 1947 and 1964 (see Figure 1.4 in Chapter 1).

By sampling randomly from this distribution, we assume that the properties of the random variation that we have observed in the past will persist in the future. By properties, we mean characteristics such as the mean, variation, and shape of the distribution. Why do we sample randomly, instead of using the correct sequence of growth rates between 1947 and 1964? We cannot use the exact sequence of growth rates because there is no guarantee that the environmental factors between 1947 and 1964 will repeat themselves in exactly the same order in the future. It would be a very strong assumption (meaning very likely to be wrong) to assume that they would. Instead we make a generalization based on this observed distribution: We assume that the distribution of growth rates in the future (their mean, variation, etc.) will remain the same as the observed distribution, even if the growth rates do not repeat themselves in the same order as in the period from 1947 to 1964.

Of course, even if we sampled randomly, the set of growth rates we chose will probably be different from what actually will happen in the future. To account for the inherent uncertainty of the future growth rates, we do this many times. We randomly select a set of growth rates for, say, 20 years, and estimate the population's future with these 20 growth rates. This gives one possible future for the population. Then we select another 20 random numbers, and repeat the process. By undertaking repeated trials we may predict the population size into the future, accounting for the effects of the environment on the population. In order to get a representative sample of possible futures of the population, we have to repeat this hundreds of times. This procedure is most easily implemented on a computer (actually, it is next to impossible to do without a computer).

The procedure may be further generalized by sampling the growth rates from a statistical distribution that has the same properties as the variations that have been observed in the past. For example, we may sample the distribution known as the normal distribution, with the same mean and standard deviation as the observed distribution. This approach has the advantage of recognizing that values of R more extreme than those observed in the past are possible in the future. For instance, if we observed the population for 100 years instead of 17, perhaps there would be a year with a growth rate of 0.8 or 1.4.

Before we proceed, we need to define some terms we will use frequently in describing stochastic models. A time series of population abundances is called a population trajectory. When we estimated the population's future with 20 randomly selected growth rates, we produced a population trajectory. Each trial or iteration that produces a population trajectory is called a replication. Finally, running the model with many replications is called a stochastic simulation. A deterministic simulation produces a single population trajectory without any variation in model parameters.

The Muskox population in 1936 was begun with 31 animals. Applying our current knowledge of the population, we can make predictions for the population over the period before regular sampling, between 1936 and 1948. The results of 1,000 trials for the Muskox population are shown in Figure 2.6. This figure shows, for each year, the average expected size (dashed curve), plus and minus one standard deviation (vertical lines), and the maximum and minimum values recorded for that year (triangles). These statistics (mean, standard deviation, minimum and maximum) are calculated over the 1000 replications (trials) of simulated population growth. The five observed values for the Muskox population size made between 1936 and 1948 are also shown (black circles). The model includes both demographic and temporal environmental variation. The growth rate, R, is 1.148, the survival rate, s, is 0.921, and the standard deviation in the growth rate is 0.075 (based on the observed variation in Figure 2.5).

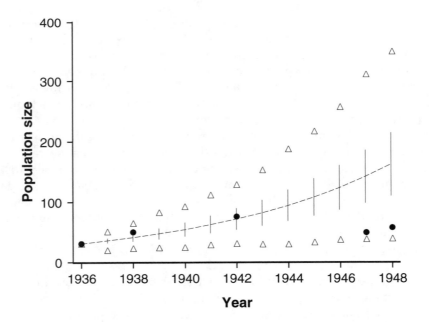

Figure 2.6. The size of the Nunivak Island Muskox population, based on 1,000 replications.

The population grew much as could have been expected between 1936 and 1942. However, between 1942 and 1947, the true population was markedly reduced, compared to growth in periods before and after that interval. In 1947, the population numbered just 49 animals. The observers suggested that the losses were due to groups of animals wandering onto sea ice during winter and being lost, other accidental deaths, and shooting. The observed population size in 1947 was within the limits of what could have been expected, once the random variations due to demographic and environmental uncertainty were included in the prediction.

The variation in the predicted abundance increases as time goes on (Figure 2.6). Our predictions become less and less certain, the further into the future we make predictions. This characteristic is a general result common to all predictions that include uncertainty. It makes good intuitive sense. One can be more certain of predictions that are made in the short term. Long-term judgements are subject to many more uncertain events, and the bounds on our expectations must increase, the further into the future that we make projections.

It is possible to construct a quasi-extinction risk curve based on the projections that are summarized in Figure 2.6. One simply records the smallest size to which the population falls during each trial. There will be 1,000 such records from 1,000 trials. These numbers are then used to create a cumulative frequency histogram. The frequencies, rescaled between 0 and 1, and plotted against population size, become the risk curve (Figure 2.7a).

If one collects the smallest value recorded at any time during each trial, the risk curve has a specific meaning. It tells us the chance that the population will fall below the specified threshold at least once during the period over which predictions are made.

Of equal interest is the creation of explosion risk curves. It is possible to construct a curve representing the chances that the population will be greater than or equal to a specified threshold population size. The procedure is much the same. One records the largest size to which the population rises during each trial. These numbers are used to create a cumulative frequency histogram. The frequencies, rescaled between 0 and 1, and plotted against population size become the explosion risk curve (Figure 2.7b). Extinction risk curves are useful for questions related to the likely lower bound of a population. Explosion risk curves are useful for questions related to the likely upper bound of a population.

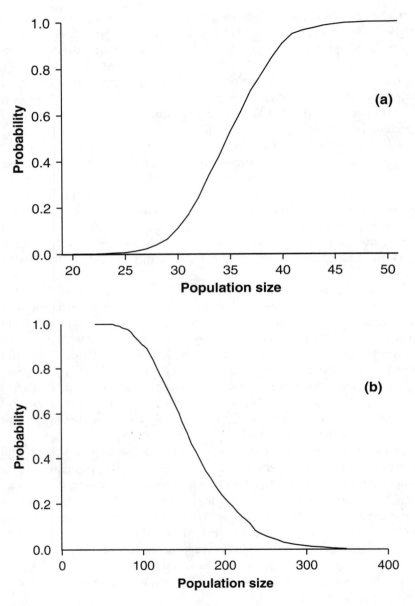

Figure 2.7. Risk curves for the Nunivak Island Muskox population for the 12-year period between 1936 and 1948: (a) quasi-extinction risk curve and (b) quasi-explosion risk curve, for the population based on an initial size of 31 individuals.

2.2.3.2 Spatial variation

The environment varies in space as well as in time. Changes in environmental conditions are related to distance. Two butterflies living in an oak forest in New York are more likely to experience the same kind of weather from day to day than are butterflies that live on opposite sides of the continent. Anyone who has dived in the ocean will have noticed a smooth transition from light to darkness with increasing depth. If survival or fecundity depend on environmental conditions, then they too will vary in space in response to the variation in environmental conditions.

One way of looking at spatial variation in the environment is to think of it as your ability to predict the conditions in some other place, knowing the conditions where you are. It is not possible to predict exactly the rainfall at one location, knowing the current rainfall at another location. The degree of reliability in the prediction from one place to another will depend, at least in part, on how far the two points are apart. The association between the recorded values of an environmental variable at different places is termed spatial correlation.

Spatial variation may also be thought of as the variation in environmental conditions between spatially separate patches of habitat, the different conditions experienced by each of several populations. Many species consist of an assemblage of populations that occur in more or less discrete patches of habitat. We can ignore the differences in the environment experienced by these populations only if these patches are identical in composition and close enough that they experience the same environmental conditions. In most real populations, at least one of these conditions will be violated. All of the populations will experience some environmental changes in common (such as the average summer temperature) and some will experience local environmental changes uniquely in a given patch (such as the local water hole drying out). The pattern of change in local population size in response to unique environmental conditions can have profound effects on our expectations of future population sizes. The interactions between these processes and the role of migration of individuals between patches will be explored more fully in Chapter 6.

2.2.3.3 Catastrophes

Catastrophes are extremes of environmental variation, including natural events such as floods, fires, and droughts. Any environmental change that has a relatively large effect on the survival or fecundity of individuals in a population compared to the normal year to year fluctuations may be considered a catastrophe. Thus, it is somewhat arbitrary to single out and label such environmental conditions as extreme. The category is useful only insofar as some ecological processes are driven by relatively infrequent, catastrophic events. In many ecosystems, disturbances such as fire, windstorms, or snowstorms are an important or even the dominant ecological process determining the structure and composition of populations and communities. Often, we may know quite a lot about the characteristics of these events such as their average frequency and the distribution of intensity of the events. With field data it is possible to specify the effects of catastrophes on the parameters of a population. If so, then there will be better understanding of the relationship between the population and the environment incorporated in the expressions that we write.

Explicit modeling of unique catastrophic events may even be essential for circumstances in which species are specially adapted to the effects of the catastrophe. For example, seeds of many plant species in the genus *Acacia* require a fire to germinate. In the absence of fire, adults produce seeds that mostly fall to the forest floor and remain dormant. Fires stimulate the germination of dormant seeds and kill adults, which have life spans of 10 to 100 years in the absence of fires. Thus, recruitment of new individuals into the population occurs in pulses following the fires that stimulate germination and eliminate adults. Fecundity is a binary condition: either there is none (in years without fire) or most seeds in the soil-stored seed bank germinate (in years with fire). Such dynamics could only be modeled by writing expressions that include the chance of a fire.

2.3 Parameter and model uncertainty

2.3.1 Parameter Uncertainty

In all of the above discussion, we have assumed that the quantities obtained from field observation including mean survival, fecundity, growth rate, and the variation in these parameters, are known exactly. Effectively, we have assumed that the observed variation in population parameters comes from sources including demographic and environmental variation. Anyone who has attempted to measure the simplest parameter more than once under field conditions knows that this is a false assumption. All measurements involve error.

Parameter uncertainty is the variation in our estimate of a parameter that is due to the precision and accuracy of the measurement protocol. The assumption that sampling error is absent is particularly suspect when data are limited. Smaller samples are subject to relatively large sampling errors. If sampling variation is included in a model, projected variability will be much larger than in the true population. The Muskox of Nunivak Island provide an example. Aerial census techniques were used to estimate population size between 1949 and 1968. These data were used to calculate all of the parameters in the examples used up to the present. However, between 1964 and 1968, independent estimates were made based on ground samples (Table 2.2).

Table 2.2. Counts (from ground samples) and estimates (from aerial samples) of the total population size of Muskox on Nunivak Island.

Year	Count	Estimate	Bias (Count/Estimate)
1965	532	514	1.035
1966	610	569	1.072
1967	700	651	1.075
1968	750	714	1.050

After Spencer and Lensink (1970).

The aerial "estimates" of population size were consistently lower than the ground-based counts. If we assume that the counts are correct (and there is no absolute guarantee of that), then the estimates were consistently biased, but the magnitude of the bias varied from year to year, from 3.5% to 7.5% (Table 2.2). Bias may be defined as systematic error, the difference between the true value and the value to which the mean of the measurements converges as more measurements are taken. Precision is the repeatability of a measurement made under the same conditions. Unfortunately, we do not have any estimates of Muskox population size made in the same year using the same method. Such data would allow us to quantify the precision of the population estimates.

Often, subjective judgment is involved in the choice of a method for measuring a parameter. Similarly, judgment may be made in assuming a correspondence between one variable and another. For example, we may observe that rainfall varies by 10% each year, and assume that population growth varies by the same amount. Even more subtle is the assumption that the levels of variation that we have observed in the past will persist in the future. There is nothing wrong with such judgments; often they are unavoidable. However, it is wrong to ignore the uncertainty inherent in such judgments.

2.3.2 Model Uncertainty

The structure of a model relates the parameters to the dependent variable, in this case future population size. If our ideas concerning the population's dynamics and ecology are wrong, or if we have not been careful in translating our ideas into equations, our predictions may be astray. Uncertainty concerning the form and structure of the expressions we use to describe the population is known as model uncertainty. Thus, even if the parameters that describe the dynamics of a population were known exactly, and the variation associated with each parameter was decomposed into demographic uncertainty, environmental uncertainty and measurement error, we could still make mistakes in predicting future population size.

Model structure is a simplification of the real world. It represents a compromise between available data and understanding, and the kinds of questions that we need to answer. It is difficult to know the degree of simplification that is both tractable and adequate to the task at hand, but that is not so simple that it misses some important ecological processes. Competing model structures may provide as good, or almost as good, explanations of past observations as one another, but generate quite different expectations. The only way to deal with model uncertainty is to compare predictions of models with different structures and (if they make different predictions) to analyze the models in detail to understand which assumptions led to the differences. Such an analysis may guide further field observations or experiments to decide which model structure is more realistic.

2.3.3 Sensitivity Analysis

Both parameter uncertainty and model uncertainty may be explored using a process known as sensitivity analysis. Sensitivity analysis measures the change in a model's predictions in response to changes in the values of parameters, or to changes in the model structure. To illustrate sensitivity analysis, consider the model in which a population's growth rate is related to several environmental variables. For example, variation in the growth rate of a population of Shrews (*Crocidura russula*) that inhabit suburban gardens in Switzerland is related to weather variables by

$$\Delta R \approx 0.73 \cdot P - 0.78 \cdot S + 0.50 \cdot T_s - 0.83 \cdot T_w$$

where P is mean monthly precipitation in spring (m), S is winter snow fall (m), and T is average monthly mean temperature (°C) in summer (T_s), and winter (T_w). We know that summer rain averages about one meter and that winter snow fall averages about the same value. The coefficients for the two parameters are similar. Thus, the growth rate will be equally sensitive to variations in snow fall and rainfall. The coefficients for temperature are about the same magnitude. However, the values for temperature vary more (they are around 10°C in summer and around 5°C in winter), so that R is effectively more sensitive to variations in temperature. A 10% increase in summer temperature will increase R by 0.5, whereas a 10% increase in snow depth will decrease R by only 0.08.

The object of sensitivity analysis is to tell us which parameters are important and which are not. If a 10% change in a parameter results in a small change in the dependent variable (say, less than 1%), the model is insensitive to the parameter. If the change in the dependent variable is large (more than 10%), then the model is highly sensitive to the parameter. Such information is useful because it may guide the direction of research effort. It is more important to eliminate measurement errors from parameters to which our predictions are sensitive than to eliminate it from parameters that contribute little to our predictions.

Sensitivity analysis may also be used to explore alternative model structures. For example, our model for the growth rate of a population above may have the best explanatory power in a statistical sense. However, our biological intuition may tell us that the following model is likely to be a better predictor of future population growth:

$$\Delta R \approx 0.15 \cdot P \cdot T_s - 0.7 \cdot S$$

In this version, P and T_s are multiplied because we treat the effect of rainfall and summer temperature as an interaction. We may fix the parameter values and explore the consequences for predictions of one model versus the other. In some cases, the model structure will make little difference to expected outcomes. In other cases, it will make an important difference. If the latter is true, it would be advisable to perform experiments or acquire more data to discriminate between the competing models. If the acquisition of data or experimental results are impossible, then predictions may be made with both models, and the most extreme upper and lower bounds may be used to place limits on the predictions. In this way, predictions can incorporate model uncertainty that is not reducible without further field work.

The above example was based on a statistical relationship between population growth rate and environmental variables. Sensitivity analysis may be based on other variables as well. It is important to evaluate both the deterministic and the probabilistic components of a prediction. Thus, the dependent variable against which we judge model sensitivity may be the risk of population extinction within a specified period of time, or the risk of the population increasing above some specified upper bound. The independent variables would be model parameters and their variation. If an increase in a parameter (say, average growth rate or the standard deviation of growth rate) results in more than a 10% increase in risk, then the model may be considered to be sensitive to that parameter. We will further explore this type of sensitivity analysis in the exercises of this section.

Sensitivity analysis is perhaps one of the most important tools in quantitative population ecology. It allows us to explore the consequences of what we believe to be true (in terms of the model parameters and their ranges). It provides a measure of the importance of parameters and model assumptions. It may be used to place bounds on predictions that subsume both model and parameter uncertainty, providing a relatively complete picture of the reliability of a prediction.

2.4 Ambiguity and ignorance

In natural resource management, rare and unexpected events may be termed "surprises" (see Hilborn 1987). Ignorance leads to surprise. It may result from unawareness of unexpected events, or from false knowledge or false judgments. That does not mean that surprise itself is rare, only that each event is essentially unexpected. It includes anything we do not expect, anything that is unaccounted for by our model or by our intuition.

Some surprises are avoidable because the ignorance they spring from may be reducible. That is, it may be amenable to study or learning. One may be ignorant of a process or a predictable outcome, but could overcome that ignorance by learning or research if the information or the methods of study were available. There are direct and indirect costs of such ignorance. For example, ignorance of past experiments or observations may lead to the tacit acceptance of hypothetical results, without empirical testing. It may cause disciplines such as wildlife management to loose credibility with people with a vested interest in wildlife.

Other surprises may be unavoidable. We may be unaware that we are unable to make predictions accurately, if the structure of the system were to change. That is, we would be faced with novel circumstances. For example, the demographers studying the human population as recently as 60 years ago predicted that the population size would be 3 billion by the end of the century. It will probably be over 6 billion. They were wrong by a factor of two, in part because of unavoidable surprises. They could not have foreseen the decrease in mortality caused by the invention of antibiotics, or the increase (albeit temporary) in food production as a result of widespread use of pesticides.

Uncertainty may arise from disagreement, even amongst scientists interpreting the same information. Interpretations are colored by a person's technical background, expertise, and understanding of the problem. Things are further complicated by the fact that people, decision makers and scientists included, frequently hold direct or indirect stakes in the outcome of a question. Judgments are influenced by motivational bias.

Linguistic imprecision may be responsible for important components of uncertainty. The statement "the population is not threatened by what we plan to do" is ill-specified. To interpret it, we need more information. Would the statement be true if the probability of decline of the population to half its current size was 10% in the presence of the impact, and 2% in the absence of the impact? Even so, many more specifics are needed.

A quantity is called well specified when there is a single true value that is measurable, at least in theory. The test for clarity of specification of a problem is whether it can be unambiguously defined, given a description. For example, the phrase "Provide a management plan that results in an acceptable risk of decline of a population" is an ambiguous request. Risks include both a probability and a time frame, so one must first ask, What is the time horizon over which one wishes to estimate the risk? Secondly, the term "acceptable" is undefined. The concept of an acceptable risk will vary depending on the magnitude of the decline, whom you ask, and what it is they have to gain or lose by various management strategies. Thus, ambiguity in the specification of a problem may create kinds of uncertainty that are beyond any kind of quantitative or qualitative analysis, and it may be resolved only by political or social processes. We will explore these concepts further in the final chapter of the book.

2.5 Additional topics

2.5.1 Time to Extinction

The quasi-extinction risk curves we examined focus on probability of falling below certain levels anytime during a fixed interval of time (thus we call them "interval" risk curves). For example, we used a 12-year period or interval in the Muskox example (Figure 2.7a). A different way to express the results of the simulation is to keep a record of the time it takes each replication of the simulation to become extinct (or fall below a critical threshold abundance). If we ran the simulation for a long time and recorded the year of extinction for each of the 1000 replications, we could use these data to construct a time-to-extinction curve, the same way we used minimum abundances to construct risk curves. A time-to-extinction curve (Figure 2.8) gives the probability that the population will have gone extinct by the time a given number of years (x-axis) have passed.

60   Chapter 2 Variation

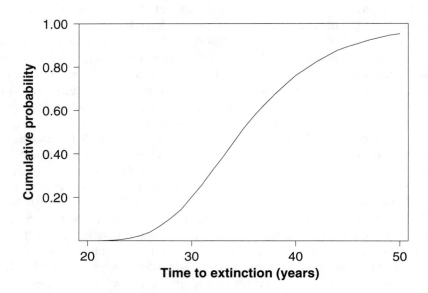

Figure 2.8. Time-to-extinction curve (the number of years that will pass before a hypothetical population falls below a fixed threshold).

Note that the curve looks similar to the quasi-extinction risk curve, but it has a very different meaning. In this case (Figure 2.8), the x-axis gives the number of years, and the threshold of extinction is fixed. In the case of the risk curve above (Figure 2.7a), the x-axis gives the threshold, and the time interval is fixed. In this book we will mostly use the risk curves, but briefly come back to the time-to-extinction curve in a later chapter.

2.5.2 Estimating Variation

Very often, estimates of population size through time are used to calculate parameters for population growth models. In Figure 2.6, the standard deviation representing variation around the mean population size was predicted by a simple population model that included both demographic stochasticity (see Section 2.2.2) and environmental variation (see Section 2.2.3). In this model, the environmental variability was modeled by a population growth rate that varied randomly from one year to the next. The amount of variation in the growth rate is measured by its standard deviation. In this case, the standard deviation was 0.075. This estimate was based on the observed, year-to-year variation in growth displayed in Figure 2.5. In other words, the number 0.075 is the standard deviation of the 17 yearly growth rates from 1947 to 1964.

There are some problems with this approach. The observed variation in growth rate (Figure 2.5) has several sources, including environmental change between years, demographic stochasticity, and sampling (measurement) error. Even if the environment was constant, demographic variation and sampling error would ensure that the rate of change in the size of the population changes (or appears to change). When estimating the standard deviation of the growth rate (which we used in the model that produced Figure 2.6), we assumed all of the variation is due to environmental change.

This assumption may be reasonable if the population is large (so that demographic variation is negligible) and the size of the population is known with a high degree of reliability (so that sampling error is negligible). In other circumstances, to assume that all observed variation in growth rates in due to the environment alone will overestimate the true variation in the population.

We should subtract the sampling variance and the demographic variance from the total variance estimate. The difference would be variance due to the environment. In general, this is difficult to do correctly and it is a topic of ongoing, active research. In the meantime, assuming that all variation is due to the environment generally will tend to result in estimates of extinction and explosion probabilities that are too high. It is important to remember this fact when interpreting the results of a study, and to explore the consequences for the results of relatively small values for environmental variation.

2.6 Exercises

Before you begin this set of exercises, you need to install the program RAMAS EcoLab, if you have not yet done it. Read the Appendix at the end of the book to install RAMAS EcoLab on your computer.

Exercise 2.1: Accounting For Demographic Stochasticity

In this exercise, you will predict the change in population size of the Muskox population between 1936 and 1937, accounting for demographic stochasticity. For this exercise you will need to choose uniform random numbers. Some calculators give a uniform random number every time you press a key. If you have one of these, you can use it (skip "Step 0" and go to "Step 1"; you will need two such numbers for each repetition of this step). If you don't have such a calculator, you can use RAMAS EcoLab (see "Step 0").

Step 0. Start RAMAS EcoLab, and select "Random numbers," which is a program that produces random numbers. The program will display two uniform random numbers (between 0 and 1) on the screen. To get another pair of random numbers, click the "Random" button. (To quit, select "Exit" under the File menu, or press Alt-X.)

Step 1. The Muskox population consisted of 31 individuals in 1936. Write down this number ($N = 31$) on a piece of paper. Repeat the following steps 31 times, once for each Muskox on Nunivak island in 1936. For each repetition, use a new pair of random numbers.

Step 1.1. Use the first random number to decide if the animal produces an offspring or not. If the first random number is *less than* the fecundity value ($f = 0.227$), then *increase* N by 1, otherwise leave it as it was.

Step 1.2. Use the second number to decide if the animal survives or dies. If the second random number is *greater than* the survival rate ($s = 0.921$), then *decrease* N by 1, otherwise leave it as it was.

Step 2. After repeating the above steps 31 times, record the final N. This is your estimate of the Muskox population size for 1937.

Step 3. Repeat Steps 1 and 2 four times, for a total of five trials. You will have 5 estimates for the Muskox population size for 1937. Comment on the amount of variation among the results of the five trials.

Exercise 2.2: Building a Model of Muskox

In this exercise, you will use RAMAS EcoLab to build and analyze a stochastic model of Muskox on Nunivak island.

Step 1. Start RAMAS EcoLab, and select the program "Population Growth (single population models)" by clicking on its icon. See the Appendix at the end of the book for an overview of RAMAS EcoLab. For on-line help, press F1, double click on "Getting started," and then on "Using RAMAS EcoLab." You can also press F1 anytime to get help about the particular window (or, dialog box) you are in at that time. To erase all parameters and start a new model, select "New" under the Model menu (or, press Ctrl-N).

Step 2. From the Model menu, select **General information** and type in appropriate title and comments (which should include your name if you are going to submit your results for assessment).

Enter the following parameters of the model.

Replications:	0
Duration:	12

Setting replications to 0 is a convenient way of making the program run a deterministic simulation, even if the standard deviation of the growth rate is greater than zero. Note that the last parameter of this window, whether to use demographic stochasticity, is ignored (it is dimmed and is not available for editing). This is because when the number of replications is specified as 0, the program assumes a deterministic simulation. This parameter is ignored because it is relevant only for stochastic models.

After editing the screen, click the "OK" button. (Note: Don't click "Cancel" or press (Esc) to close an input window, unless you want to undo the changes you have made in this window.) Next, select **Population** (under the Model menu). Recall that the Muskox population on Nunivak Island began in 1936 with 31 individuals and had an average growth rate of 1.148. Based on these, enter the following parameters in this screen.

Initial abundance:	**31**
Growth rate (R):	**1.148**

The parameter "Standard deviation of R" is not available for editing because we will first run a deterministic simulation, in which standard deviation will not be used. Similarly, "Survival rate (s)" is used only to model demographic stochasticity, so it is also ignored by the program when the simulation is deterministic.

For this exercise, you can ignore the last two parameters in this window (density dependence and carrying capacity); we will discuss density dependence in a later chapter. The default selection for "Density dependence type" is "Exponential," which refers to exponential growth with no density dependence. The last parameter is ignored because it is related to other types of density dependence. When finished, click "OK" and press (Ctrl-S) to save the model in a file.

Step 3. Select **Run** from the Simulation menu to start a simulation. The simulation will run for 12 time steps; you will see "Simulation complete" at the bottom of the screen when it's finished. For a deterministic simulation, this will be quite quick. Close the simulation window.

Step 4. Select "Trajectory summary" from the Results menu. Describe the trajectory you see. What is the final population size?

Step 5. Close the trajectory summary window. Select **General information** and change "Replications" to 100 by typing the number. Next, click the little box next to "Use demographic stochasticity" This will add demographic stochasticity to the model. The parameters should now be as follows:

Replications:	**100**
Duration:	**12**
☑ Use demographic stochasticity *(checked)*	

Click the "OK" button and select **Population** (again, under the Model menu). Recall that the survival rate of the Muskox population was 0.921 and that the observed standard deviation in the growth rate was 0.075. Based on these, enter the following parameters in this screen.

Initial abundance:	**31**
Growth rate (R):	**1.148**
Survival rate (s):	**0.921**
Standard deviation of R:	**0.075**

Click "OK," and select **Run** to start a simulation. While this stochastic simulation is running, after the first five replications, the program will display each population trajectory it produces (the program cannot display the population trajectories produced by the first five replications, because it uses them to scale the graph). Describe the trajectories in comparison with the deterministic trajectory. Do any of these trajectories look similar to the deterministic trajectory? What is the cause of the difference?

Step 6. After the simulation is completed, close the simulation window and save the model by pressing (Ctrl-S). Then, select "Trajectory summary." You will see an exponentially increasing population trajectory. Describe the trajectory summary. What is the range of final population sizes? You can try to read the range from the graph, or if you want to be more precise, you can see the results as a table of numbers. To do this, click on second button from left ("show numbers") on top of the window. The first column shows the time step, the others show five numbers that summarize the abundance for each time step: (1) minimum, (2) mean – standard deviation, (3) mean, (4) mean + standard deviation, and (5) maximum.

Step 7. Select "Extinction/Decline" from the Results menu. What is the risk of decline to 31 individuals based on this curve?

It might be difficult to read the precise value of the risk from the screen plot. Do the following to record this number precisely:

Click the "Show numbers" button, and scroll down the window to where you see "31" in the first column. Record the probability that corresponds to this threshold level.

Exercise 2.3: Constructing Risk Curves

In this exercise you will construct an interval decline risk curve based on the Muskox model. If you have exited the program after the previous exercise, first open the file you saved at Step 6 in Exercise 2.2 (press Ctrl-O) and choose the file you saved). If you did not save the previous model, then enter the parameters as described in Step 5 of the previous exercise.

Step 1. In the next step we will generate single trajectories. To prepare for this, select **General information**, and change "Replications" to 1. Also, change "Duration" to 5. Make sure that "Use demographic stochasticity" is checked. Click OK. (Note: If you want to save the model in this exercise, use "Save as" and give the file a different name, so you keep the original file.)

Step 2. Generate a single random trajectory based on the model in Exercise 2.2. To do this, run the model and display the trajectory summary as a table of numbers (see Step 6 in the previous exercise). Record the smallest value that the population trajectory ever reached during time steps 1 through 5 of this single replication. (Note: Ignore time step 0, for which the abundance is always 31.)

Step 3. Repeat Step 2 a total of 20 times.

Step 4. You now have 20 minimum population sizes from 20 runs. Sort these in increasing order, and use the table layout below to generate frequencies from the records of minimum population sizes. In the first column of the table, write the population sizes you have in increasing order. You are likely to get some population sizes more than once. Write these down only once. You will most likely use only some of the rows in this table. In the second column, write how many of your numbers is the population size in column one. In the third column, cumulate the numbers of the second column (see Table 2.1). In the fourth column, calculate probabilities by dividing the cumulative frequencies (third column) by the number of trials (20). Note that this table is similar to Table 2.1, but your numbers will be different because you have only 20 trials or replications, whereas Table 2.1 was constructed based on 10,000 trials.

Step 5. Plot the probabilities against population size in Figure 2.9.

Population size (N_c)	Number of runs that reached a size $\leq N_c$	Cumulative number of runs (that reached a size $\leq N_c$)	Probability of decline to N_c

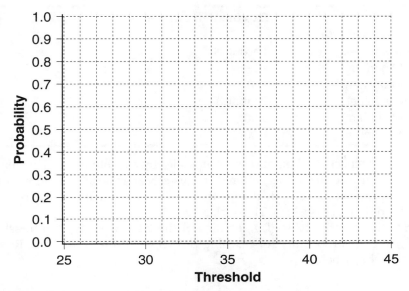

Figure 2.9.

Exercise 2.4: Sensitivity Analysis

In this exercise, we will use the Muskox model from Exercise 2.2 to analyze the sensitivity of quasi-explosion probability to model parameters. Our aim is to decide what parameter is more important in this particular model in determining the probability that the Muskox population will increase to 150 individuals. You might consider this probability a measure of the success of the reintroduction project: Assume that the project is regarded as successful if the Muskox population reaches 150 individuals within 12 years.

Step 1. Load the stochastic Muskox model you saved in Step 6 of Exercise 2.2. In this exercise, we will call this model the "standard model." View the "Explosion/Increase" curve. Record the threshold and the probability of increasing to 150 individuals.

It might be difficult to read the precise value of the probability from the screen plot. Do the following to record this number precisely. (This procedure can also be used for "Extinction/Decline"; it is similar to, but more detailed than the one in Exercise 2.2.)

Click the "Show numbers" button, and scroll down the window to where you see "150" in the first column. Record the probability that corresponds to this threshold level. If "150" is not in this table, then click the third button on top of the window ("scale"). You will see a window with various plotting parameters (the exact numbers may be different in your simulation).

68 Chapter 2 Variation

Title:	**Explosion/Increase**
☑ Autoscale *(checked)*	
X-Axis Label:	**Threshold**
Minimum:	46
Maximum:	456
Y-Axis Label:	**Probability**
Minimum:	0.00
Maximum:	1.00

First, uncheck the box next to "Autoscale" by clicking on it. (This makes the program use the values entered in this screen instead of automatically rescaling the axes.) Second, change the maximum value of the *x*-axis to the threshold (in this case, 150). Third, click OK.

Scroll down the table. The last line of the table will give the threshold (150), and the probability of reaching or exceeding that threshold. Record this probability below.

```
Probability of increasing to 150 =
```

Step 2. Create eight new models based on the standard model. For each model, increase or decrease one of the four parameters of the model (see below) by 10%, and keep all the other parameters the same as the standard model. Note that there are some restrictions. For example, the survival rate (s) cannot be less than 0 or greater than 1. And the initial abundance must be an integer. Make necessary adjustments or approximations for these parameters. Save each model in a separate file. Record the low and high value of parameters, and filenames that contain them.

Initial abundance:	**31**
Growth rate (R):	**1.148**
Survival rate (s):	**0.921**
Standard deviation of R:	**0.075**

Parameter:	*low value and filename*	*high value and filename*
Initial abundance		
Growth rate (R)		
Survival rate (s)		
Stand. deviation of R		

Step 3. Run stochastic simulations with the eight models you created in the previous step. After each simulation, view the quasi-explosion results, and record the probability that the Muskox population will increase to 150 individuals within the next 12 years. Record the results in the table below.

Probability of increasing to 150

Parameter:	with high value	with low value	difference
Initial abundance			
Growth rate (R)			
Survival rate (s)			
Stand. deviation of R			

Step 4. For each parameter, subtract the probability with low value from the probability with high value of the parameter. Discuss the results.

(a) In which direction did each parameter affect the result? (In other words, does higher value of the parameter mean higher or lower probability?)

(b) Which parameter affected the outcome most, when the change was 10%? What should this result tell about field studies which attempt to estimate these parameters, or about future projects similar to this one?

Note that sensitivity of the result to ±10% of survival rate, or growth rate, or its standard deviation can be interpreted in terms of accuracy in the estimation of these parameters, or in terms of the value of these parameters in other places where a similar project will be implemented. However, sensitivity of the result to ±10% of initial abundance cannot be interpreted in terms of accuracy: It is probably not very difficult to count 31 animals. However, it might be interpreted in terms of the effect of the initial number of individuals on the success of the project.

2.7 Further reading

McCoy, E. D. 1995. The costs of ignorance. *Conservation Biology* 9:473-474.

Morgan, M. G. and M. Henrion. 1990. *Uncertainty: A guide to dealing with uncertainty in quantitative risk and policy analysis.* Cambridge University Press, Cambridge.

Shaffer, M. L. 1987. Minimum viable populations: coping with uncertainty. In M. E. Soulé (Ed.). *Viable populations for conservation* (pp. 69–86). Cambridge University Press, Cambridge.

Chapter 3
Population Regulation

3.1 Introduction

The Muskox population we studied in Chapters 1 and 2 was growing at a rate of about 14.8% per year; this growth continued for about 30 years. What would happen if this population actually continued to grow for another 30 years? If you repeat the exercise, starting from the final abundance of 700 Muskox and ran the model for another 30 years, you would see that the final abundance would be about 44,000 muskox. Another 30 years, and it would be 2.7 million! Obviously, this is not what happens in nature; this species has been around for millions of years. How does exponential growth come to an end?

As we discussed in Chapter 1, exponential growth happens under favorable environmental conditions. Sooner or later, the environment will not be favorable; for example, a series of severe winters will occur during which there will be few calves born. This will set back the population abundance. For other species, the reason may be too much water (floods), too little water (drought), or any of the factors we discussed in Chapter 2 that cause fluctuations in the environment.

It is possible that extrinsic factors (such as climate) will remain favorable for a long time, even as they fluctuate. When that happens, the population will continue growing and become crowded. The resources that are available to the population will have to be shared among an ever-increasing number of individuals. These resources include water, food (for animals), nutrients and sunlight (for plants), and space. As we discussed in the beginning of Chapter 1, individuals of the same species share the same niche, i.e., they have similar requirements for such resources, which is why crowding forces each individual to get a smaller share of the available resources. In addition to depleting its food resources, a growing population may poison the environment with its own wastes, and attract predators and diseases.

3.2 Effects of crowding

The reactions of species to overcrowding vary greatly. In some species, crowding causes increased dispersal; many individuals leave the population and look for less crowded places. For example, in cyclic populations of voles, dispersal rate increases as the population approaches its peak densities for the cycle. We will discuss dispersal in more detail in the chapter about metapopulation dynamics.

3.2.1 Increased Mortality

In some species, increased dispersal cannot compensate for increased densities. In other words, dispersal alone cannot reduce the densities to levels where there are no crowding effects. This is especially true for species that disperse only passively. For example, dispersal in many plants occurs as the dispersal of seeds by wind, animals, etc. In such cases, increased density may mean that the mortality rate will increase. As more seedlings share limited space, water, or other resources, more of them will die. In an experiment, seeds of *Cakile edentula* (which is an annual plant that lives on sand-dunes) were sown at densities of 1 to 200 per 400 cm^2. The survival rate of these seedlings was inversely related to the initial sowing density (Figure 3.1). In this experiment, seed survival was defined as the proportion that produced mature fruits, so it actually included both survival and reproduction.

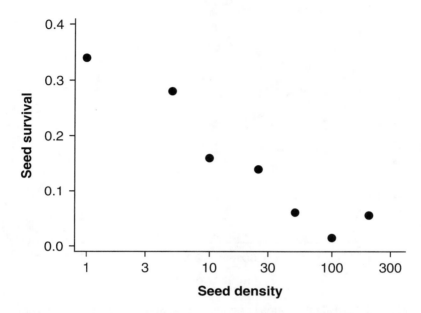

Figure 3.1. Effect of density on seed survival in *Cakile edentula*, an annual plant. Data from Keddy (1981).

3.2.2 Decreased Reproduction

When individuals get fewer resources, they may reproduce less, or even cease to reproduce altogether. For example, the clutch size in many bird species depends on the available resources. An example of the effect of crowding on fecundity is the change in the fledging rate in a Great Tit *Parus major* population in Oxford, England, shown in Figure 3.2. The number of fledglings (i.e., the number of young birds leaving the nest) per breeding bird is used as a measure of fecundity. The figure shows that as the size of breeding population increases from about 40 to about 90 birds, the fecundity decreases from about 4.5 fledglings per bird to about 2.5. Despite the large amount of variation (originating perhaps from fluctuations caused by environmental factors), the effect of crowding on fecundity is quite evident. (We will return to this graph later in this chapter, when we caution about using such graphs to quantify the density dependence relationships.)

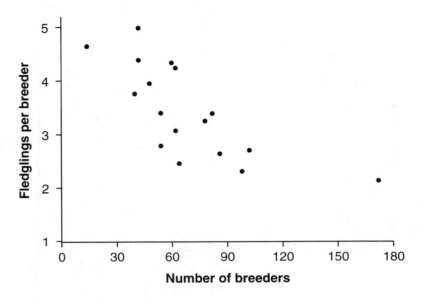

Figure 3.2. Effect of density on fecundity: Great Tit *Parus major*. Data from Lack (1966).

3.2.3 Self-thinning

When density-dependent mortality takes its toll and the density of the population declines, the remaining individuals are better off. This is especially evident in plants; the decline in population size due to density-dependent mortality is usually more than compensated by the increased size of surviving plants, and as a result, the total biomass actually increases. In several cases, this compensation of mortality with growth in the size of individuals follows a very specific and regular pattern, called a *self-thinning curve*. This is often plotted on log-log scales, with the density of plants in the horizontal axis and mean weight of plants in the vertical axis (see Figure 3.3).

There is considerable (although not unequivocal) evidence that self-thinning curves, when plotted on log-log scales, show a slope of −1.5. If a stand of plants are sown at sufficiently high density, the change in mean weight and density through time follows a slope of −1.5.

Effects of crowding 75

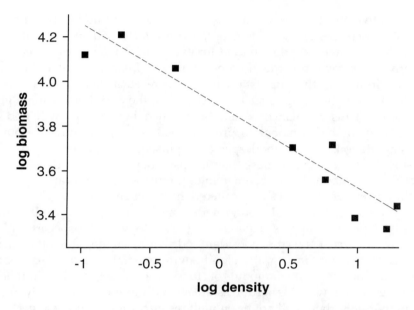

Figure 3.3. Self-thinning curve for White Birch *Betula pubescens* (after Verwijst 1989). Density is number of plants per unit area and biomass is measured as the mean weight of a plant. Each point corresponds to a stand of trees.

3.2.4 Territories

A territory is an area of the habitat defended by an individual or pair, and from which other individuals are excluded. Examples of territorial species include many bird species. Spotted owls, for example, hold territories that may be 10 to 40 km^2 and that are defended by a pair of breeding owls against other spotted owls. When density increases in a population of a territorial species, one (or more) of several things may happen: territories may become packed more tightly, territories may get smaller, some animals may be pushed to less than optimal habitat, and, last but not least, a larger number of individuals may be excluded from all territories. Adult birds without territories are often called "floaters." Floaters do not breed; thus as population density increases, the average fecundity of all individuals declines, even though the average fecundity of breeding individuals may remain the same.

3.3 Types of density dependence

The negative effects of crowding discussed above will build up as population size increases. At low densities, these effects will be negligible; for example, there will be enough food for everyone. As densities increase, the negative effects will become more pronounced. Unless stopped by unfavorable environment, the population will soon reach a density at which the negative effects build up to such an extent that the population cannot grow anymore. Thus a balance would be reached between growth and the capacity of the environment to provide food, space, and other resources. These limitations, be they food, space, self-poisoning, or predators, are called *density-dependent factors* because their intensity depends on the density of the affected population. The phenomenon of population growth rate depending on the current population size is called density dependence.

How exactly the population growth comes to a stop as a result of these density dependent factors depends on the ecology of the species and the limiting resource. Let us first consider food shortage and ignore other factors such as emigration, immigration, and environmental variation. As population size increases, the amount of food resources for each individual decreases. If the available resources are shared more or less equally among the individuals, there will not be enough resources for anybody at very high densities. Such a process of sharing leads to a type of intraspecific (*within* the species) competition called *scramble* competition. In contrast, *contest* competition occurs when resources are shared unequally, and there are always some individuals who get enough resources to survive and reproduce.

There are other ways in which competition occurs differently in different species or for different resources. For example, competition may be indirect, through the sharing of food, but without actual physical contact or confrontation between the competitors. Or it may be direct, with actual fighting over the limited resources. We will not concentrate on such differences; what matters from the modeling point of view is whether resources are shared equally or unequally.

An example of scramble competition might be competition for food among fish larvae (newly hatched fish). If there are very few individuals, almost all of them may survive. If the density of larvae is very high, none of them will get enough to survive. This is an extreme example of scramble competition, in which the total number of survivors is less when there are more individuals to start with. This process is also called *worsening returns* because as density increases, conditions for the whole population, not just the average individual, get worse.

An example of contest competition is competition for territories, in which the winner gets all of the territory (and can not only survive but also reproduce) while the loser gets none. In this case, no matter how many compete for the limited number of territories, there are always some breeding individuals. The process of *contest* competition is also called diminishing returns. As the number of competitors increase, the proportion that can find a territory diminishes, and the total reproduction (e.g., total number of offspring from all territories) increases more and more slowly, but it does not decline.

3.3.1 Scramble Competition

The two types of competition we discussed are important because they have quite different effects on the dynamics of the population and its risk of extinction. To add effects of density in our models, we need to decide what causes density dependence, and add appropriate equations to our model. There are many ways to write equations to add density dependence to the models we have been using in earlier chapters. We will discuss some of these equations later in this chapter. For now, we don't need to worry about the specifics of the equations; we will instead concentrate on their general characteristics. We will compare various types of density dependence with two types of graphical representation.

One graphs abundance at the next time step (next year, for instance) as a function of abundance now. Such a function is shown in Figure 3.4. The type of representation in this figure is called a *recruitment curve*, or a *replacement curve*. It shows what the population size will be next year (the y-axis), given the population size this year (the x-axis). In scramble competition, this function is humped and the right end of the curve is declining; in other words, for large population sizes, the population size at the next time step, $N(t+1)$, is a *declining* function of the population size at this time step, $N(t)$. The curve always starts from the origin, because if the population size is zero, then it will also be zero next year (assuming no immigration). See Section 3.8.1 for the equation that we used for scramble competition.

The dotted (45°) line shows exact replacement, where $N(t)$ equals $N(t+1)$. The level of abundance at which this line and the replacement curve (continuous curve) intersect is labeled as K on the graph. The replacement curve (continuous line) is above the exact replacement line at the left part of the graph, to the left of the point of intersection of the two. When the current population size is in this region (N less than K), $N(t+1)$ is greater than $N(t)$, which means the population will grow. When the current population size is greater than K ($N>K$), then $N(t+1)$ is less than $N(t)$, which means the population will decline. When the current population size is equal to K, then $N(t+1)$ is equal to $N(t)$, which means the population size will remain the same. So,

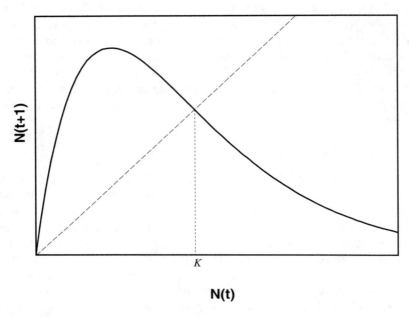

Figure 3.4. Replacement curve for scramble competition.

the population tends to grow when it's small, and it tends to decline when it's large. This property makes this type of density dependence *stabilizing*. Since it pushes the population up from lower abundances and down from higher abundances, it stabilizes the population size around a certain level. In mathematical terms, this level is called an *equilibrium point* because of this stabilizing effect, and is represented by the parameter K. In a density-dependent model, the equilibrium point can be described in biological terms as the *carrying capacity* of the environment for the population, i.e., the population size above which the population tends to decline. Another related term we introduce is *regulation*. A population is said to be regulated when its density is kept around an equilibrium point by density-dependent factors.

We can use a replacement curve to make a deterministic prediction of the population's future. All we need is the initial population size. The top graph (A) in Figure 3.5 is a replacement curve, and the bottom one (B) is the population trajectory based on this curve.

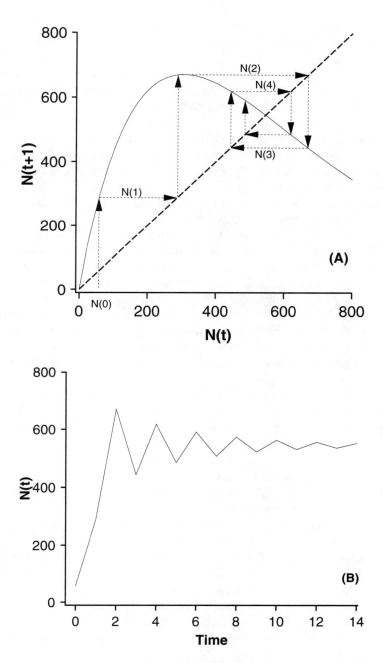

Figure 3.5. Predicting the population trajectory (B), based on the replacement curve (A).

First, find the initial population size, indicated by $N(0)$ in graph (A). Second, find $N(0)$ in graph (B). In the second graph the x-axis is time and y-axis is the population size. For $x = N(0)$, the curve in graph (A) gives $N(1)$, the population size in the next time step—in other words, the y-value that corresponds to $x = N(0)$. We read this from the graph as about 300, and plot it in graph (B) for $t = 1$. To predict $N(2)$ from $N(1)$, we find the y-value predicted by the replacement curve for $x = N(1)$. An easy way to do this is to extend a horizontal line from the curve at $x = N(0)$, $y = N(1)$ to the exact replacement line (the dashed, 45° line), and a vertical line from the 45° line back to the curve. Now the y-value is $N(2)$, which we plot in graph (B). For the remaining time steps, we repeat the same process: drawing a horizontal line from the curve to the 45° line, and a vertical line from the 45° line back to the curve, and plotting the y-value in graph (B).

The reason the replacement curve is also called the recruitment curve has to do with its historic association with fisheries biology. *Recruitment* in fisheries refers to the natural increase in the harvestable portion of the population (fish above a certain size) by growth of smaller (e.g., newly hatched) fish. Typically, only a small fraction of eggs become recruits. The larger the number of eggs, the more intense the competition between the newly hatched individuals, and the smaller their chances of survival. The points in Figure 3.6 show the recruitment data for Bluegill Sunfish *Lepomis macrochirus*, a popular freshwater sport fish widely introduced throughout the temperate world.

The bluegill data were used to generate the replacement curve (dotted), using one of the equations for modeling scramble type density dependence: the Ricker equation, developed by the fisheries biologist W.E. Ricker. The point at origin was added to assist the model fitting (Figure 3.6). This is justified because zero eggs would give zero recruits. The fit of the data to the model is remarkably good for this type of study, which perhaps has something to do with the fact that Ricker (1975) developed his equation while studying bluegill population dynamics.

Another type of graph by which we can visualize a density dependence function is shown in Figure 3.7, which gives the growth rate of the population as a function of abundance. The growth rate in this case is calculated as the ratio of the population size in the next time step to the population size now, $N(t+1)/N(t)$. The exact replacement line (which was a 45° line in the previous figure) in this figure is a horizontal line at growth rate equal to 1.0. The population grows when growth rate is greater than 1.0 (i.e., when the curve is above the line), and declines when the growth rate is less than 1.0 (when the curve is below the line).

Types of density dependence 81

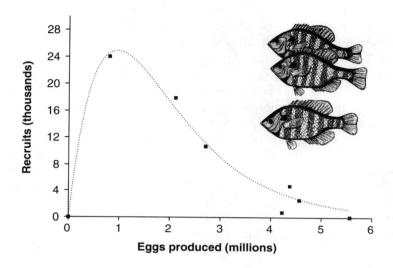

Figure 3.6. Density dependence in Bluegill Sunfish. After Ferson et al. (1991).

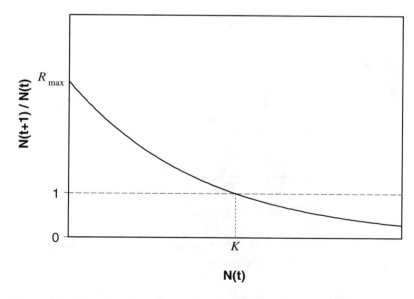

Figure 3.7. Density dependence in growth rate for scramble competition.

The two types of graphs that we used to represent density dependence are quite similar, they give the same information. There is one important piece of information that we can get more easily from the second type of graph. The *y*-intercept of the curve (the point at which the curve intercepts the vertical axis) marks the maximum rate of growth (R_{max}), which happens as the population size approaches zero, where the effects of crowding do not affect the population. This is the rate of exponential growth we discussed in Chapter 1; in other words, it tells how fast this population will start to grow, if it grows free of density-dependent effects. Because the densities will be small at the beginning, the change in population size through time will initially look like exponential growth. This initial phase of density-dependent population growth is seen in the first few days of an experiment by the Russian biologist G.F. Gause in the 1930s (Figure 3.8). Gause started the experiment with 2 *Paramecium aurelia* (a protozoan species). After the first few days, the growth rate of the *Paramecium* population started to decline, and after day 12 or so, the population size fluctuated around 550.

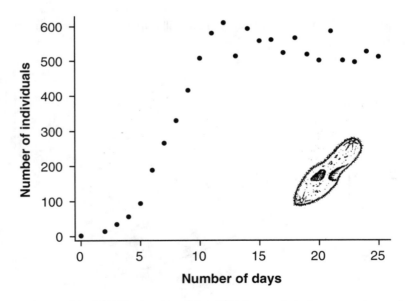

Figure 3.8. Growth of a *Paramecium aurelia* population starting from 2 individuals. Data from Gause (1934).

In Chapter 1, we used a logarithmic scale for abundance in order to demonstrate the exponential nature of the growth of the population of Muskox. If the abundance graph through time is linear in the logarithmic scale, this means that the population is growing exponentially. When the population growth starts exponentially, and then slows down, as it did in the case of the *Paramecium* population in Gause's experiments, then the logarithmic graph will be linear at the beginning, and curve down to a horizontal line. In Figure 3.9, we show the same data as in the previous figure, but with the population size (*y*-axis) in logarithmic scale.

Note also that the fluctuations of the population after it reaches the equilibrium do not seem to be as large in Figure 3.9 as in the original figure, due to the logarithmic scale. One must be careful in interpreting graphs in logarithmic scale, which tend to play down the importance of variation and error. In Exercises 3.1 and 3.2, we will model the growth of this *Paramecium aurelia* population using RAMAS EcoLab.

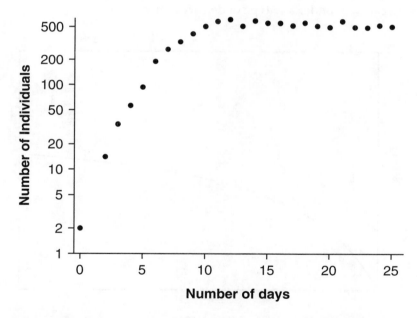

Figure 3.9. Growth of a *Paramecium aurelia* population starting from 2 individuals. Data from Gause (1934). Note the logarithmic scale for the number of individuals.

3.3.2 Contest Competition

If the available resources are shared unequally so that some individuals always receive enough resources for survival and reproduction at the expense of other individuals, there will always be reproducing individuals in the population. As we mentioned above, this will be the case in populations of strongly territorial species, in which the number of territories does not change much even though the number of individuals seeking territories may change a lot. This process of diminishing returns leads to contest competition, which is represented by the replacement curve in Figure 3.10. If you compare this figure with the replacement curve for scramble competition (Figure 3.4), you will see that the major difference is in the right side of the curve, at high population densities. Whereas the curve for scramble competition is humped, and declines at high densities, the curve for contest competition reaches a certain level and remains there. The similarity is that, in both cases, if the population size is above the carrying capacity, it will decline in the next time step. However, under contest competition, no matter how high the population density is, the population in the next time step will not be below the carrying capacity, assuming a constant environment. If there is environmental variability, the population may decline below the carrying capacity under any type of density dependence.

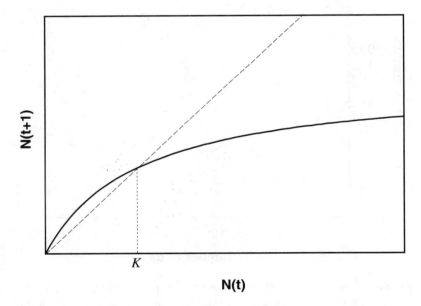

Figure 3.10. Replacement curve for contest competition.

The curve shown in Figure 3.10 is a specific type of density dependence function know as the Beverton-Holt function (see Section 3.8.1 for the equation we used for this function), based on Beverton and Holt (1957).

3.3.3 Ceiling Model

There are several other density dependence models that will give the general characteristics of contest competition. One of the simplest of these is the ceiling model. The replacement curve for ceiling density dependence is shown in Figure 3.11. Note that this is similar to Figure 3.10 in that the right-hand side of the curve is flat rather than declining.

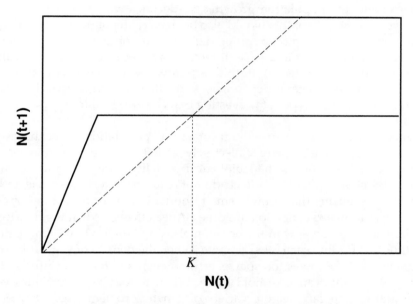

Figure 3.11. Replacement curve for the ceiling model of density dependence.

In the ceiling model, the population grows exponentially until it reaches the carrying capacity (for example until all territories are occupied), and then remains at that level until a population decline takes it below this level. If the population grows above the carrying capacity (by immigration, for example), it declines to the carrying capacity by the next time step. In this case the carrying capacity acts as a *population ceiling*, above which the population cannot increase. This is somewhat different from the types of density dependence function we studied earlier, in which the population density *equilibrated around* the carrying capacity. In contrast, the average population size in a ceiling model may be well below the carrying capacity, because the population may be pushed below this level by environmental variation, but cannot increase above it. Because of the difference between ceiling type of density dependence and the contest type based on the Beverton-Holt function, our program includes "Ceiling" as an additional choice of density dependence type, even though it may result from contest competition. We will explore these differences in Exercise 3.3.

3.3.4 Allee Effects

The three types of density dependence we explored so far share a common characteristic: the population growth rate declines with *increasing* density. However, the reverse of this would also be density-dependent: If population growth rate declines with *decreasing* density, the population growth is still density-dependent, but in a very different way (because of this, it is called inverse density dependence). In this section we will discuss natural factors that cause inverse density dependence, and their consequences. In a later section, we'll discuss types of harvesting that can also lead to inverse density dependence.

Factors that cause the growth rate of a small population to decline, as the population gets smaller, are collectively called Allee effects (named for Warder C. Allee who studied biological sociality and cooperation; Allee 1931; Allee et al. 1949). Allee effects do not result from a single cause; rather several mechanisms that draw a small population away from the carrying capacity and toward extinction are called Allee effects. When the density of whales becomes very low in the ocean, males and females have a more difficult time just finding each other to mate. When the density of vegetation on a mountain slope becomes too sparse, erosion begins to take away the soil so even fewer plants can take hold there. When a population becomes very small, inbreeding can create a variety of genetic problems (see Chapter 2). Whenever a lower abundance means a lower chance of survival or reproduction for those individuals that remain, Allee effects may occur. Similar Allee effects are also observed in plant populations. The number of seeds per plant in small populations of *Banksia goodii* (a shrub) were less than the number of seeds per plant in larger populations. This species is pollinated by mammals and birds, and the reason for lower fertility in smaller populations was thought to be decreased number of pollinator visits (Lamont et al. 1993).

We mentioned earlier that a crowded population may attract predators, leading to density-dependent predation mortality. However, if the density of the predator in the area is constant, then predation may cause inverse density dependence. As the number of individuals of the prey species in the same area increases, the damage the predators do will be distributed among a larger number of prey, and the proportion of the population lost to predation will be lower. A decline in predation rate as a result of increased concentration of prey is also called *predator saturation*. This is one of the reasons that some species of birds form flocks to roost, or breed in colonies.

Breeding in colonies may have other social benefits, too. For example, Birkhead (1977) found that when the colonies of the Common Guillemot (a marine bird species) were dense, the hatching of eggs were more synchronized. This made sure that there were more adults around the colony when most of the chicks were still at nest, and decreased mortality from predators such as gulls. The result was that, as shown in Figure 3.12, the breeding success (measured as the proportion of nests with at least one fledgling) was higher in denser colonies. Allee effects will occur when a decline in the population of such colonially nesting species causes declines in breeding rates, which cause further declines in the population size.

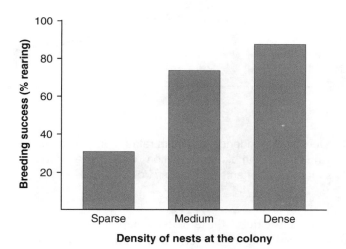

Figure 3.12. Percentage of nests rearing at least one chick to fledgling in colonies of the Common Guillemot (*Uria aalge*). Data from Birkhead (1977).

In a density dependence graph, Allee effects are represented by a curve that declines as abundance gets smaller. The density dependence curve (growth rate as a function of abundance) for scramble competition that we studied earlier is repeated as curve (a) of Figure 3.13. Other curves show how this density dependence relationship changes with the addition of increasingly stronger Allee effects. Note that these are *not* replacement (recruitment) curves; they show the relation between growth rate and abundance (not abundance next year as a function of abundance this year).

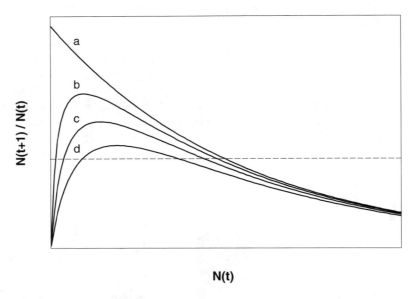

Figure 3.13. Density dependence in growth rate for scramble competition (curve a) and scramble competition with three levels of (increasingly strong) Allee effects (curves b, c, d).

An important characteristic of these density dependence curves is that the growth rate is below 1.0 at the left end of the curve, which means the population will decline if its abundance is low. This means that if a population declines to a low level by chance, then Allee effects can pull it down even further. Clearly, such phenomena can dramatically influence the risks of extinction. It also means that when Allee effects are present, the sigmoidal growth from low abundances to the carrying capacity (that we saw in Figure 3.8, for example) is not always possible. If the population starts at a low abundance, it may go extinct even before it starts growing.

The models that can be implemented in RAMAS EcoLab cannot explicitly incorporate Allee effects (although other, research-oriented RAMAS programs can). One way to incorporate Allee effects in a model is to adjust the extinction threshold. Suppose you know that Allee effects must be important for a particular species you are modeling, once its population falls below, say, 100 individuals. After running a stochastic simulation, you can view the quasi-extinction risk curve and record the risk of falling below 100 individuals. Remember, the quasi-extinction curves in RAMAS EcoLab are "interval" extinction curves, which refer to falling below a threshold *at least once during the simulated time interval*. If you assume that the relevant risk is the risk of falling below 100 individuals, this means that you don't need to worry about how realistic the model is below this abundance. In a way, you can assume the population to be extinct once it falls below this level.

3.3.5 The Concept of Carrying Capacity

In our models, we used the term carrying capacity as equivalent to the parameter K of the models, which is the level of abundance above which the population tends to decline. Defined this way, carrying capacity is obviously a characteristic specific to a population. Different populations of the same species may have very different carrying capacities, as a result of the different amounts of resources (food) or space (territories) available to them, as well as the abundances of competitor and predator species. In summary, we do not make a distinction among carrying capacities set by different types of factors. The carrying capacity is simply an abstraction that crudely summarizes the interactions of a particular population with its environment and describes the capacity of the environment to support a population, in units of the number of individuals supported.

As any abstraction, the concept of carrying capacity has its limitations. One of these is that carrying capacity, if determined by such factors as food and predators, must fluctuate as these factors change over time. The models we use in this chapter assume that carrying capacity remains constant, and whatever fluctuations there are in the environment affect the growth rate of the population. There are models in which carrying capacity is also allowed to vary, but the distinction between how environmental variation is added to carrying capacity versus to growth rates is beyond the scope of this book (see "Adding environmental variation" below).

Another limitation of our definition of carrying capacity is that to some it may imply (wrongly) that all populations are expected to end up at this abundance. There are at least four reasons why this may not be true. We already discussed two of these and we'll discuss two others later in this chapter: (1) The major reason is that environmental fluctuations will push the size of the population up and down, and therefore even populations under the effect of strong density dependence may never reach this level. Later in this chapter, we will discuss methods of adding this important factor to our density dependence models and demonstrate it with computer exercises. (2) When the density dependence model is ceiling type (see above), the dynamics of the population above and below the carrying capacity are different, and thus the population size may not stabilize at K. Moreover, even the average abundance predicted by the models may be quite different from K. (3) As we discussed in the previous section, if the density dependence model includes Allee effects, and the population size starts from a low initial abundance, the population may become extinct before it reaches K. (4) Finally, strong density dependence of the scramble kind may lead to oscillations even without any environmental fluctuation. We will discuss this later, in Section 3.5.

3.3.6 Carrying Capacity for the Human Population

As we mentioned above, the models we use in this chapter assume that the carrying capacity remains constant. The human population is one example of a population that does not fit a static concept of carrying capacity. In addition to changes in environmental conditions (for example, droughts) that may change the carrying capacity, the actions of humans themselves have greatly affected the carrying capacity of the earth for the human population. In an earlier chapter, we mentioned the unexpected effects of antibiotics and pesticides in increasing the limits to the human population. Technology and innovation have increased the carrying capacity of the earth for the human population, and many economists still believe that human ingenuity will always find answers to increased demand. From an ecological point of view, the damage to natural ecosystems in all parts of the world is an indication that the nature of the interaction of the human population with its environment is changing.

As the human population approaches its carrying capacity, its interactions with the natural environment will determine the ultimate size of the population, as well as the conditions in which humans will live. Characterizing this interaction is one of the important challenges facing applied ecologists. As we discussed in Chapter 1, the impact of humans on the environment is a function of the number of people, consumption rate per person, and environmental damage per unit of consumption. Changes in these variables through time and among different regions of earth make it impossible to calculate the carrying capacity of earth for the human population. Nevertheless, there have been several attempts to calculate it (see Cohen 1995), mostly based on consumption of renewable natural resources such as food and fresh water. Another factor that might play a role is the spread of infectious diseases, made easier by increased human densities and increased long-distance travel.

Limitation of population growth by either shortage of food or by diseases is an unpleasant prospect. A more optimistic scenario is a decrease in fertility rates or consumption rates as a result of social and economic factors. Increased national wealth or economic activity (measured by gross national product, GNP) has been associated with decreased fertility in many industrial counties. Of course, population size is only one factor that determines the human impact on the environment; increased wealth is also associated with increased consumption. Among social factors, Pulliam and Haddad (1994) found that fertility was negatively correlated with contraceptive use and with education level.

3.4 Assumptions of density-dependent models

The assumptions of a density-dependent model are similar to those we discussed in Chapter 1, with one exception. Instead of assuming that density will remain low enough to have no effect on population dynamics (assumption 3 in Chapter 1), we assume that increased density will cause a decline in the population growth rate in such a way that the growth rate will reach 1.0 when population density (N) is K (the carrying capacity), and will cause even further decrease in growth rate if $N>K$.

Regarding variability (assumptions 1 and 2 in Chapter 1), we assume that there is no variability in carrying capacity from year to year (see Section 3.3.5). In Exercise 3.2, we will demonstrate the effect of incorporating demographic stochasticity in a deterministic density dependence model, and in Section 3.7, we will discuss adding environmental variation.

All the other assumptions are the same: We assume a single, panmictic population (assumption 5), in which the composition of individuals with respect to age, size, sex, genetic properties, and others remains constant (assumption 4), and in which processes of birth and death can be approximated by pulses of reproduction and mortality (assumption 6).

3.5 Cycles and chaos

Strong density dependence functions of the scramble type can induce wild population fluctuations even in models without any environmental variation. This phenomenon is called *deterministic chaos* and has been the subject of much interest in biology and physics as well as other fields during the last several years. In its simplest form, cycles caused by density dependence proceed as follows. An initial high density (much above carrying capacity) causes a population crash. This happens when the density dependence curve declines very fast at high densities (to the right of the hump in Figure 3.4), so that a very high initial density causes the density in the next time step to be much below the carrying capacity. In the next time step, the population starts with this very low density and, as a result, grows quickly. Such cycling is illustrated in Figure 3.14, which shows the average densities of seedling of crucifer *Erophila vernau* (an annual plant) in two types of plots: plots with high initial density of seedlings, and those with low initial density.

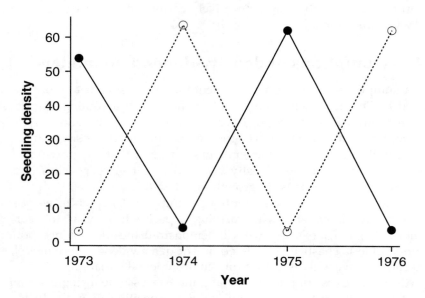

Figure 3.14. Mean density of seedlings of an annual plant in plots with initial high and low densities. Data from Symonides et al. (1986).

Symonides et al. (1986) found that most plots with intermediate densities stayed at intermediate density in the following year, whereas most of the plots with high or low densities alternated from one year to the next, as shown in the figure. This type of dynamics arises with scramble type density dependence and high growth rates. If the growth rate is really high, the cycles turn into chaos. Chaotic dynamics are characterized by the fact that small changes in the initial conditions (i.e., $N(0)$, the initial abundance) result in much larger changes in the rest of the population's trajectory. We will demonstrate these kinds of dynamics in Exercise 3.4.

3.6 Harvesting and density dependence

Earlier in this chapter, we discussed Allee effects that cause inverse density dependence. A second, human-caused way a population may experience inverse density dependence is through harvesting.

In Chapters 1 and 2, we discussed the effects of harvest or removals on the populations of Blue Whales and Muskox. Harvesting a natural population can be done with various strategies. Here we will consider two simple ones: constant harvest, where a fixed *number* of individuals are taken out of the population at each time step, and constant rate, where a fixed *proportion* of individuals are harvested.

Assuming that the natural (unharvested) population itself has no density dependence (it is growing exponentially as in the examples of Chapter 1 and 2), then it will also have no density dependence when it is harvested at a constant rate. This is because the constant rate imposes a fixed mortality, which is the same as reducing the survival by a fixed amount. Because the decrease in survival is fixed, and it does not depend on density, the population growth (or decline, as the case may be with overharvesting) will be density-independent.

However, the constant harvest is a different story. Suppose you decided on a constant harvest strategy of 10 Muskox per year. If the population size is 1000, this will mean that harvesting imposes an additional 1% mortality. If the population size is 100, the additional mortality would be 10%. The lower the population size, the higher the additional mortality due to harvest, and the lower the population growth rate. This obviously introduced an inverse density dependence, because growth rate declines with declining population density.

One of the effects of this inverse density dependence is that it tends to destabilize the population's dynamics. Effectively, whenever the population is reduced by some random change in the environment or by a series of unlucky events, the proportion of the population that is removed is greater. The reduction in population size due to chance events is amplified by the removal strategy. Thus, the removal strategy will increase the magnitude of natural fluctuations and increase the risks of crossing lower population thresholds that may be unacceptable for economic, social, or ecological reasons.

The effect of constant harvest is similar to Allee effects, but it applies to the whole range of population sizes, not just to small populations. Like Allee effects, constant harvest may easily push a population to extinction, because a lower population size may result in a decline in growth rate, which causes further declines in the population size.

The reason this form of density dependence (declining growth rate at low densities) is called inverse is perhaps historical; it does not necessarily mean that the other type (declining growth rate at high densities) is more common. A related (and somewhat confusing) terminology is positive and negative feedback. The crowding effects are said to be a form of negative feedback, whereas inverse density dependence (e.g., Allee effects) is a form of positive feedback. "Positive" in this case does not refer to the rate of growth, but to the fact that plus makes more plus and minus makes more minus (low abundance leads to even lower abundance).

We will demonstrate the inverse density dependence imposed by constant harvest in Exercise 3.5, and we will model the effects of different types of harvesting in a later chapter.

3.7 Adding environmental variation

So far in this chapter, we assumed that the density dependence relationships are constant. In other words, the growth rate of a population may change from year to year as a result of changes in population density, but the nature of the function that determines these changes remains the same.

How would we add environmental variation to a density-dependent model? The density dependence function is determined by its parameters, which, for scramble or contest types discussed above, are the maximum growth rate, R_{max}, and the carrying capacity, K. Both of these parameters may be affected by environmental fluctuations. We studied in Chapter 2 an example of how growth rate of the Muskox population varied from year to year.

We can also model environmental variation in carrying capacity. This is a more complicated concept, since carrying capacity is often measured as the long-term average abundance of a population regulated by density dependence. For a territorial species, there is usually some variation among territories in terms of the quality of habitat. In years with unfavorable environmental conditions these territories may become unsuitable, and as a result the number of territories may fluctuate from year to year. This process can be modeled by a randomly varying carrying capacity, with methods similar to those we used in Chapter 2 for randomly varying growth rates.

In this book, we will model all environmental variation as if it affects the growth rate and assume that carrying capacity is not subject to environmental variability. We will, however, study a different type of change in carrying capacities. This is not a random change, but a deterministic change or a trend in carrying capacity of a population through time. By this, we mean a change that results in consecutive increases (or decreases) in the carrying capacity. An example of such a decrease (a negative trend) is habitat loss. An example of an increasing carrying capacity might be increase in habitat quality for a forest-dwelling species as the forest grows (assuming it is not logged). Many species, including the Northern Spotted Owl, depend on older forests. For such species, a forested habitat will become more suitable, and the carrying capacity will increase, if the forest is left undisturbed for a long time. We will study how to model effects of such changes in Chapter 6.

Having decided to model environmental variation in growth rate, we still need to decide how exactly to do this. We cannot use the same method we did in Chapter 2, since now the growth rate depends on density as well as on random environmental fluctuations. In Chapter 2, we simply sampled a growth rate from a predetermined random distribution. The distribution can either be that of growth rates observed in the past, or some more general statistical distribution. In this chapter, we modify this procedure a little. Instead of selecting the growth rate from the same distribution every time step of a simulation (for example, every year), we select it from similar but different distributions each year. These distributions are similar because they have the same shape and the same standard deviation. They are different because they have different means. The means are different because they are determined by the abundance or density of the population. In other words, at a given time step, the population has a certain abundance (density), and the density dependence relationship determines the *average* growth rate as a function of this abundance. The actual growth rate for that time step is then sampled from a random distribution with this average that was determined by density dependence (see Figure 3.15).

3.8 Additional topics

3.8.1 Equations

There are several different ways of writing equations for the types of density dependence we discussed in this chapter. One of the earliest equations used was the logistic equation, which was originally developed for continuous time (differential equation) models. Another is the Ricker equation (expressed in discrete time) that we mentioned earlier with respect to the data on bluegill. For modeling scramble type density dependence in RAMAS EcoLab, we use a discrete form of the logistic equation that is mathematically equivalent to the Ricker equation. The growth rate at time t is calculated as a function of the density at time t, $N(t)$, using the following equation.

$$R(t) = R_{max}^{\left(1 - \frac{N(t)}{K}\right)}$$

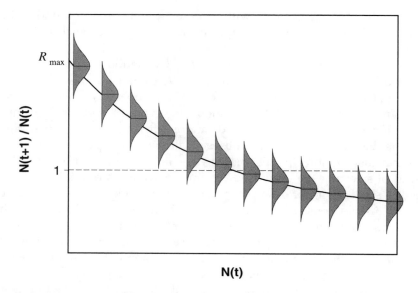

Figure 3.15. Conceptual model for adding environmental variation to a density dependence relationship. The density dependence function determines the average growth rate (as a function of the current abundance). The growth rate is then sampled from a random distribution with this average.

where R_{max} is the maximum growth rate and K is the carrying capacity. In this chapter we use R_{max} instead of R because the growth rate depends on density, and its average does not make much sense. R_{max} is the average growth rate at low population densities, where the effects of density dependence are so weak that they can be ignored.

When the population size $N(t)$ is small, the exponent $\left(1 - \frac{N(t)}{K}\right)$ is close to 1.0, and the growth rate $R(t)$ is close to R_{max}. When $N(t)$ is equal to the carrying capacity, then the exponent is zero and $R(t)$ is equal to 1.0. When the population size is above the carrying capacity, then the exponent is negative and $R(t)$ is less than 1.0, i.e., the population declines. Combining this formula with the population growth formula we used at the beginning of Chapter 1, the equation for the abundance at the next time step is

$$N(t+1) = N(t) \cdot R_{max}^{\left(1 - \frac{N(t)}{K}\right)}$$

For contest-type density dependence, we use an equation equivalent to the Beverton-Holt equation used in fisheries management. With this equation, the growth function becomes

$$N(t+1) = N(t) \cdot \frac{R_{max} \cdot K}{R_{max} \cdot N(t) - N(t) + K}$$

This formula looks a bit more complicated, but you can easily verify for yourself that when $N(t)$ is small, the population grows exponentially; when $N(t)=K$, the growth rate is 1.0 [i.e., $N(t+1)=N(t)$]; and when $N(t)>K$, the population declines [i.e., $N(t+1)<N(t)$].

For ceiling type, we use a much simpler formula:

$$N(t+1) = \min(R_{max} \cdot N(t), K),$$

which says that the population size next time step is the minimum of the two numbers, $[R_{max} \cdot N(t)]$ and K.

3.8.2 Estimating Density Dependence Parameters

Developing a density-dependent population model involves three steps: determining the presence of density-dependent population regulation, determining the type of density dependence, and estimating model parameters.

Merely observing a growth rate above 1.0 at low population densities does not justify using scramble- or contest-type density dependence. You need other evidence that shows that the population *is* regulated by these types of density dependence. This is because the observed growth rates are affected by factors other than density, such as stochasticity. Populations that are not regulated by scramble- or contest-type density dependence (for example, those that are only subject to ceiling-type density dependence) will also frequently experience periods of positive growth, some of which will coincide with low population sizes. If you model such a population with a scramble- or contest-type density dependence, you may underestimate extinction risks because of the stabilizing effect of these functions. For a discussion of complexities inherent in detecting density dependence, refer to Hassell (1986), Hassell et al. (1989), Dempster (1983), Solow (1990), and Walters (1985).

Estimating the parameters of a density dependence model may be a complicated problem. Simply using the maximum growth rate ever observed may overestimate the strength of density dependence, since this observed value might have been the result of factors other than the low density of the population or the lack of competition. One method is to use time-series data to fit one of the density dependence functions. Suppose that you have a time series of population estimates, N_1, N_2, N_3, etc.; and for each time step t, you calculate the growth rate as $R_t = N_{t+1} / N_t$. The density dependence curve discussed above suggests that, to estimate R_{max}, you need to make a regression of R_t on N_t, and use the y-intercept as the estimate of R_{max} (assuming it is a declining function). However, note that N_t appears both in the dependent and in the independent variable of the regression! This implies that if N_t are measured with error (as they are in most cases), or are subject to other (stochastic) factors, then the estimate of the slope of the regression and, hence the estimate of R_{max}, will be biased. In particular, you may detect density dependence even though it does not exist. Figure 3.2 in Section 3.2.2 is another example in which the two variables (fledglings per breeder and number of breeders) are not statistically independent.

There are no simple solutions to this problem. In some cases, averaging the two growth rates around the time step t (N_{t+1}/N_t and N_t/N_{t-1}), and regressing this variable on N_t might be helpful. If you use a geometric average, the dependent variable becomes $R_t = \sqrt{N_{t+1} / N_{t-1}}$. Note that this does not contain the independent variable N_t, thus no statistical bias would be introduced to the estimation.

3.9 Exercises

Exercise 3.1: Gause's Experiment with *Paramecium*

In this exercise, we will try to model the growth of the *Paramecium* population in Gause's experiment, using RAMAS EcoLab.

Step 1. Start RAMAS EcoLab, and select the program for single-population models. We will start with a deterministic model, which means we will set the number of replications to zero (this is how we tell the program to run a deterministic simulation). The duration should be 25 days. The program doesn't need to know that we are measuring time steps in days, but we need to keep that in mind when selecting the parameters of the model. We enter the parameters of the density dependence function in the **Population** window (under the Model menu), which is also where we enter the initial population size, 2. In addition to the initial population size, there are three other parameters you need to specify here. Let's begin with the type of density dependence. If you go to the box for "Density dependence type" and click on the little arrow to the right of the box, you will see a list from which you can select by clicking. For this model select "Scramble." Another parameter is the growth rate, which shows the rate of growth at low population sizes. This parameter should be greater than 1. We will estimate this parameter with trial and error. Enter an initial guess, say 1.5. Note that this is the growth rate *per day*, since our time step is one day. A third parameter is the carrying capacity, K, which is equivalent to the equilibrium population size. Estimate this number from the average number of individuals after day 12 in Figure 3.8, and enter it (as an integer) in the line for carrying capacity. After entering the parameters, click "Apply."

You can see the two types of density dependence graphs we discussed above for your model by clicking the "Display" button in the **Population** window. A menu will give you two choices. "Density dependence in R" gives growth rate, $N(t+1)/N(t)$, as a function of density, $N(t)$. "Replacement curve" (recruitment curve) gives $N(t+1)$ as a function of $N(t)$. Select by clicking. When finished, close the window, and click "Cancel" to close the menu.

Step 2. Select **Run** to start a simulation. When the simulation is over, select Trajectory summary from the Results menu.

How well does the predicted trajectory fit the observed one in Figure 3.8? Does it reach the carrying capacity around day 12 as the experimental population did? If it reached K earlier, this means the estimated growth rate is too fast. If it reached K later than 12 days, it means the estimated growth rate is too low.

Step 3. Repeat Step 2, by changing the growth rate according to the results, until you find the R that fits the observation it terms of how fast the population grows.

Compare the dynamics of the population as predicted by your model and the real observations in Figure 3.8, especially for the second half of the graph after day 12.

Save your model by pressing Ctrl-S, and typing a name for it (such as Gause1). The program will add the appropriate file extension, and will also save the results if they are available.

Step 4. Change the initial abundance to 800, and run another simulation. Describe the population trajectory.

Exercise 3.2: Adding Stochasticity to Density Dependence

In the previous exercise we have ignored variation. Adding stochasticity to a density dependent model can be done in several different ways. In this exercise, we will demonstrate one of these, demographic stochasticity.

In his experiments, Gause tried to keep the conditions in his laboratory setup stable and was probably successful in maintaining constant temperature, light, humidity, food, etc. Despite this, we see that there is considerable variation in the size of the *Paramecium* population. This variation may be due to demographic stochasticity, as we discussed in Chapter 2. To test this assertion, we can add demographic stochasticity to the model you developed in the previous exercise.

Step 1. To add demographic stochasticity to a model, you first need to change two parameters in **General information**. The number of replications must be greater than zero (say, 100), and the box for "Use demographic stochasticity" must be checked. This is the easy part. The difficult part is in specifying survival and fecundity in the **Population** window. As we discussed in Chapter 1, growth rate, R is equal to the sum of survival and fecundity, since the number of individuals in the next time step is the sum of the number of individuals that survive from this time step, plus the number of offspring they produce that survive to the next time step. When we run a deterministic simulation, as we did in the previous exercise, we don't need to know what these two rates are, as long as we know what their sum is (i.e., what the overall growth rate is). However, if we want to add demographic stochasticity, then we must also specify the survival rate and the fecundity.

In the case of *Paramecium* this is not very difficult if we assume that all reproduction was asexual (with binary fission). In this case, survival can be assumed to be zero, and growth rate will be equal to fecundity. The growth rate you found in the previous exercise by trial-and-error was probably close to 2. Since binary fission produces 2 "offspring" from one "parent," this means that the time step of one day is quite close to the generation time of *Paramecium* in this experiment. To enter this information into the model, go to the **Population** window and make sure that survival rate and the standard deviation of R are zero. Click "OK."

Step 2. Run a simulation. How do the trajectories with demographic stochasticity compare with the observed experimental result?

Before you quit the program, save the model in a new file, such as Gause2.

Exercise 3.3: Exploring Differences Between Density Dependence Types

In this exercise we demonstrate the differences in population growth when the growth rate and carrying capacity are the same, but the type of density dependence is different.

Step 1. Load the deterministic model of *Paramecium* we developed in Exercise 3.1. Make sure that demographic stochasticity is not used (i.e., the box for "Use demographic stochasticity" is clear in **General information**). Click OK. In **Population**, make sure the density dependence type is "Scramble," and change the growth rate to 10. Click OK, and run a simulation. Look at the trajectory summary. What do you observe?

Step 2. Repeat Step 1 several times by gradually increasing the growth rate from 10 to 20.

Step 3. Now change the density dependence type to "Contest," and repeat the simulation with the same growth rate that you last used. What is the difference between the trajectories?

Step 4. Change the density dependence type to "Ceiling" and repeat the simulation with the same growth rate that you last used. What is the difference between the trajectories?

Exercise 3.4: Demonstrating Chaos

Step 1. Use the following density dependence curve (Figure 3.16) to trace the trajectory of the population for 10 time steps. The first two time steps are already simulated.

Step 2. Plot a trajectory of the population (i.e., a graph of $N(t)$ versus t). What do you observe?

Step 3. Use RAMAS EcoLab to model this population. Here are some hints:

* Make a deterministic model.
* The density dependence type is determined by the shape of the curve.
* The carrying capacity is represented by an intersection on the graph.
* Initial abundance is indicated on the *x*-axis of the graph.
* R_{max} is represented by the slope of the curve close to the origin, and is pretty high. Use trial-and-error, and check by comparing the figure with the replacement curve (from "Display" in **Population**).

Step 4. Run a simulation, and look at the trajectory summary. Is the trajectory similar to what you have plotted in Step 2?

Step 5. Change the initial abundance, and run another simulation. Is the trajectory similar to what you have plotted in Step 2?

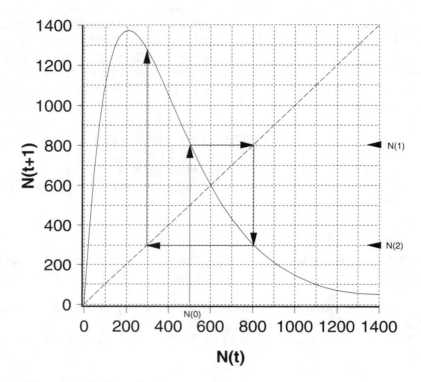

Figure 3.16. Replacement curve.

Exercise 3.5: Density Dependence and Harvesting

In this exercise, you will explore the effects of two types of harvesting in terms of their density-dependent effects. Suppose you are harvesting a fish species that has an average natural survival of 0.5 and a population that fluctuates between 2,000 and 14,000. At the end of each year, you harvest 1,000 individuals from whatever number is present in the population.

Step 1. Calculate the overall survival rate of this species (i.e., the proportion that survives both the natural causes of death and harvesting), as a function of its population size, using Table 3.1 below. For each year, first calculate the number of fish that survive the natural causes of death. Then, subtract 1,000 to simulate harvest, divide the remaining number with the number at the beginning of the year. In the table, the calculation for $N=10,000$ is provided as an example.

Step 2. Plot the overall survival rate as a function of population size in Figure 3.17.

Table 3.1.

Population size at the beginning of the year	×	Natural survival rate	=	Number that survive	−	Number to harvest	=	Number that remain	÷	Number at the beginning	=	Overall survival rate
2,000	×	0.5	=		−	1,000	=		÷	2,000	=	
4,000	×	0.5	=		−	1,000	=		÷	4,000	=	
6,000	×	0.5	=		−	1,000	=		÷	6,000	=	
8,000	×	0.5	=		−	1,000	=		÷	8,000	=	
10,000	×	0.5	=	5,000	−	1,000	=	4,000	÷	10,000	=	0.40
12,000	×	0.5	=		−	1,000	=		÷	12,000	=	
14,000	×	0.5	=		−	1,000	=		÷	14,000	=	

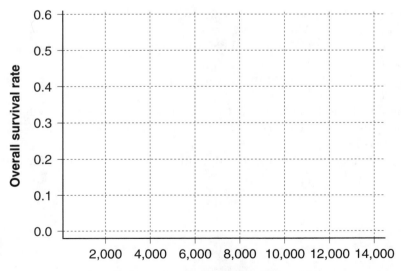

Figure 3.17.

Step 3. Now assume that the harvest strategy is to remove 40% of the fish that survive the natural causes of death (or, 20% of the total number at the beginning of the year), instead of a fixed number. Plot the overall survival rate as a function of population size (use the same figure, but a different pattern or color of curve). Compare the two curves.

Step 4. Discuss what the long-term effects of these two harvesting strategies might be on the persistence of the population.

Exercise 3.6: Density *In*dependence Graphs

In this exercise, you will find out what the two types of density dependence graphs we studied in this chapter look like, if there is *no* density dependence.

Step 1. Draw and label the axis of the two types of graphs for representing density dependence.

Step 2. To each of these two graphs, add the lines representing exact replacement.

Step 3. On each graph, draw two curves that represent (i) a population increasing exponentially at a rate of 10% per year, and (ii) a population decreasing exponentially at a rate of 10% per year.

3.10 Further reading

Cohen, J. E. 1995. Population growth and earth's human carrying capacity. *Science* 269:341–346.

Dempster, J. P. 1983. The natural control of butterflies and moths. *Biological Reviews* 58:461–481.

Hassell, M. P. 1986. Detecting density dependence. *Trends in Ecology and Evolution* 1:90–93.

Pulliam, H. R. and N. M. Haddad. 1994. Human population growth and the carrying capacity concept. *Bulletin of the Ecological Society of America* 75:141–157.

Strong, D. 1986. Density-vague population change. *Trends in Ecology and Evolution* 1:39–42.

Chapter 4
Age Structure

4.1 Introduction

In developing the exponential model of Chapter 1, we used two basic demographic parameters, survival and fecundity, to describe the growth of a population. We defined these parameters as average rates of all individuals in the population and made the assumption that individuals in a population can be considered to be identical. In particular, this assumption implies that births and deaths are independent of the ages of the individuals. This might be a reasonable assumption for some species, such as annual plants, or in cases where the differences among individuals of different ages cancel each other out, such that changes in the number of individuals of different ages do not affect the growth rate at the population level. But for many species, the age of an individual is one of its most important characteristics having strong effects on the individual's chances of survival and reproduction. For example, in most plant species, the survival chances of established individuals are much higher than those of seeds or seedlings. Newly hatched young of many marine animals such as marine turtles and most fishes also have much lower chances of survival than adults. Chances of successful breeding are also dependent on age in many species.

Differences among individuals in terms of their survival and fecundity can have important consequences for our predictions about what the population will do in the future. These differences are also important factors in terms of management options. When deciding which individuals to translocate, reintroduce, or harvest, managers need to keep in mind the survival chances and reproductive potentials of individuals. To account for these differences, we need more detailed models than those we have developed so far. One way to add more detail into models is to partition the population into age classes. Such a model is called an *age-structured model*.

Consider the case in which you have gone into the field and counted the number of birds in a population. Your results say that there are 26 birds that fledged last year (i.e., 26 birds are less than 1 year old), 16 are between 1 and 2 years old, 12 are between 2 and 3 years old, and so on. Instead of representing the population with its abundance, N, as we have done in previous chapters, we represent it by the abundances of different age classes. We can display these numbers in a vector, as follows ("vector" is a mathematical term for a column or row of numbers):

$$N(\text{this year}) = \begin{bmatrix} 26 \\ 16 \\ 12 \\ \vdots \\ \vdots \\ \vdots \end{bmatrix}, \text{ or more generally as: } N(t) = \begin{bmatrix} N_0(t) \\ N_1(t) \\ N_2(t) \\ \vdots \\ \vdots \\ N_\omega(t) \end{bmatrix}$$

Here, $N_x(t)$ denotes the number of individuals of age class x at time t. Thus age classes are denoted in subscripts; the maximum age class has the subscript ω (lowercase Greek letter omega).

In an age-structured model, the characteristics of the age classes are described by schedules of age-specific demographic parameters, instead of by the overall population growth rate (or birth and death rates). These parameters (survival, fecundity, dispersal) are conceptually similar to those in previous chapters, with one important difference; here we assign different values for different age classes.

Strictly speaking, "age" and "age class" are different concepts. Age is a continuous variable; for example, you can be 19 years, 11 months, and 29 days old. Age class is discrete; until you are 20 years old, you are in the age class of 19-year olds. Because we emphasize discrete models in this book, we will not worry much about this distinction. However, there are a few places where you need to be aware of whether a particular number or variable refers to the beginning or the end of the age class.

As before, the time step of a model can have any units, such as years or days. The same is true for age, which can be measured in any unit. In our discussion below, we will sometimes use the word "year" to make an example specific. Of course there is nothing special about this particular time step. Age structure can be defined in years, months, days, decades, or any other units that may be convenient for the particular species being modeled. An age-structured model for *Paramecium* may define age classes in terms of days, whereas one for elephants may define age classes in terms of decades. However, there are two important restrictions. First, the time step of the model and the interval of an age class must be the same. For example if a model for an elephant population defines age classes in units of one decade, it must have a decade time step. In other words, such a model will predict the population's abundance once per decade. The second restriction is that all age classes must have the same width or interval. For example, if zero-year olds comprise the first age class (i.e., the one-year interval from birth to the first birthday), then the second age class must also have an interval of one year, and consist of one-year olds (those between their first and second birthdays). The one exception to this restriction is the *composite age class*, which is described later in the chapter.

4.2 Assumptions of age-structured models

From a practical point of view, using an age-structured model implies that one can determine the age of all individuals in the population with certainty. The basic assumption of age-structured models is that the demographic characteristics of individuals are related to their age, and among individuals of the same age, there is little variation with respect to their demographic characteristics such as chance of surviving, chance of reproducing, and number of offspring they produce.

Initially, we will also assume that the population is closed, i.e., there is no immigration or emigration.

Simple age-structured populations also make additional assumptions that there is no demographic or environmental stochasticity, and that there is no density dependence. In the rest of this chapter, we will develop an age-structured model, to which we will add stochasticity (in Section 4.5) and density dependence (in the next chapter).

For simplicity, we will assume that all reproduction takes place at the same time. Such a model is called a "birth-pulse" population model, because the births take place in pulses, such as during spring breeding seasons. It is possible to construct age-structured models with the alternative assumption that births occur continuously (for example, as in humans). Such models are called "birth-flow" models; they are more complicated, but the principles behind them are the same as those behind the simpler models we will consider.

4.3 An age-structured model for the Helmeted Honeyeater

We will illustrate the basic principles and concepts involved in building an age-structured model with a hypothetical data set, based on the ecological characteristics of the Helmeted Honeyeater (*Lichenostomus melanops cassidix*), an endangered species endemic to Victoria, Australia. The Helmeted Honeyeater is a territorial bird that lives in the *Eucalyptus* swamplands. More detailed models for the Helmeted Honeyeater have been developed by McCarthy et al. (1994), Pearce et al. (1994), and Akçakaya et al. (1995).

This hypothetical data set consists of four annual censuses, conducted at the same time each year, in which all individuals in the population are counted and their ages are determined. We assume that these censuses are made right after the breeding season, which we assume is short relative to the time interval between the breeding seasons.

In addition to the basic assumptions outlined above, we will make an additional assumption to make the calculations easier. We assume that this species starts breeding at age 1 and the fertility rate does not vary with age among breeding individuals.

We also adopt the convention that individuals within their first year of life are called zero-year olds. Thus the first age class consists of zero-year-old individuals.

Suppose we collected the data in Table 4.1 by censusing this population as described above. Each row of this table corresponds to one age class, and each column corresponds to one annual census, in which all individuals in each age class were counted.

Table 4.1. Number of individuals of each age counted between 1991 and 1994 in a hypothetical population of the Helmeted Honeyeater.

	Census year			
Age	1991	1992	1993	1994
0	26	28	27	29
1	16	17	20	20
2	12	11	13	14
3	9	8	9	10
4	7	6	6	8
5	5	4	5	5
6	4	3	3	4
7	3	3	2	3
8	2	2	2	2
9	1	1	1	2
Total	85	83	88	97

4.3.1 Survival Rates

According to the table, in 1991, we censused a total of 85 individuals, which included 26 zero-year olds, 16 one-year olds, etc. Of the 26 zero-year olds we counted in 1991, 17 became one-year olds in year 1992, and the others died. The number that survived to be one-year olds (N_1) is the number we counted as zero-year olds (N_0) times the survival rate of zero-year olds (S_0):

$$N_1(1992) = N_0(1991) \cdot S_0$$

We can represent this with Figure 4.1, in which the arrows represent the survival of each age class from 1991 to 1992. We did not count any 10-year olds, which may mean that all 9-year olds died between 1991 and 1992.

Another way of expressing the equation

$$N_1(1992) = N_0(1991) \cdot S_0$$

is that the age-specific survival rate, S_x, is defined as the proportion of x-year-old individuals that survive to be $x+1$ years old one year later:

$$S_0 = N_1(1992) / N_0(1991)$$

Chapter 4 Age Structure

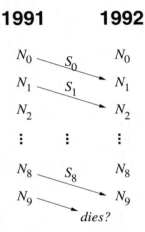

Figure 4.1. Survival of age classes from 1991 to 1992.

Using our data set, we calculate that from 1991 to 1992, the survival of zero-year olds was 17/26 = 0.654. If we make the same calculation for zero-year olds from 1992 to 1993, the survival rate is found to be 20/28 = 0.714. Finally, from 1993 to 1994, the survival rate is 20/27 = 0.741. The survival rate changes from year to year, either because of demographic stochasticity or environmental variation. To summarize the survival rate, we can use the mean and standard deviation of these three numbers. Although the correct way to calculate the mean is to use an average weighted by sample size (see Section 4.7.1), for our purposes a simple arithmetic mean is a good approximation. Averaging these three numbers, we calculate that $S_0 = 0.703$. Later, we will also use the variation in this set of numbers to add stochasticity to our model.

For the other age classes, the calculations are similar. We can write the above equation in a more general form:

$$S_x(t) = N_{x+1}(t+1) / N_x(t)$$

Using this equation for $x = 1,2,...,9$, we calculated the average survival rates for all classes as given in Table 4.2. Note that we did not count any 10-year olds, so we cannot estimate S_9, which is survival rate from age 9 to age 10.

4.3.2 Fecundities

The fecundity, F_x, is the average number of offspring (per individual of age x alive at a given time step) censused at the next time step. Note that this definition incorporates a time delay between the census of the parents and the census of the offspring. For example, the fecundity in year 1991 is the number of offspring produced in 1991 that are still alive in 1992, divided by the number of parents in 1991:

$$F(1991) = \frac{\text{Offspring alive in 1992}}{\text{Parents in 1991}}$$

Table 4.2. Age-specific survival rates based on the data in Table 4.1

Age (x)	Survival rate (S_x)
0	0.703
1	0.717
2	0.751
3	0.769
4	0.746
5	0.717
6	0.806
7	0.778
8	0.667

We can represent this with Figure 4.2, in which the solid arrows represent the fecundity of age classes and dotted arrows represent their survival from 1991 to 1992.

As stated above, we assume that this species starts breeding at age 1 and the fertility rate does not vary with age among breeding individuals. In other words, we assumed that $F_0 = 0$, and $F_1 = F_2 = F_3 = ... = F_9$. To calculate fecundity for the reproductive age classes, we divide the number of zero-year olds in the next year's census with the total number of individuals aged 1 and older (potential parents) in this year's census. For example,

$$F(1991) = \frac{28}{59} = 0.4746 \quad F(1992) = \frac{27}{55} = 0.4909 \quad F(1993) = \frac{29}{61} = 0.4754$$

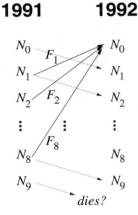

Figure 4.2. Fecundity (*F*; solid arrows) of age classes. Survivals from 1991 to 1992 are represented by dotted arrows.

The average of these three numbers is about 0.48, which we will use as the mean fecundity. See Section 4.7.1 for more information on estimating fecundities. We might also have additional information about the contribution of each age class to the production of zero-year olds. Such information allows the calculation of a different fecundity for each age class. We will demonstrate how to use such information in one of this chapter's exercises.

Given a set of age-specific fecundities, the number of zero-year olds is calculated by the formula

$$N_0(t+1) = F_0(t) N_0(t) + F_1(t) N_1(t) + F_2(t) N_2(t) + \ldots + F_\omega(t) N_\omega(t)$$

which is equivalent to

$$N_0(t+1) = \sum_{x=0}^{\omega} F_x(t) N_x(t)$$

The symbol $\sum_{x=0}^{\omega}$ means we add for all values of x from 0 to the maximum age.

4.3.3 Sex Ratio

If you are modeling both sexes, the fecundities should be in terms of "offspring (of both sexes) per individual," and the initial abundances should be the number of all individuals (male and female) in each age class. In the Helmeted Honeyeater example, we have censused and are modeling both sexes, so the fecundity is in terms of offspring per individual. In other words, when dividing this year's zero-year olds with last year's parents, we used the total numbers of offspring and parents, not daughters or adult females.

Sex ratio is the ratio of females to males in a population or in an age class. Sex ratio at birth is the ratio of daughters to sons among the offspring. In a matrix model, the sex ratio must be incorporated into the estimation of fecundities. If you are modeling only the female population, the fecundities should be in terms of "number of daughters per female," and the initial abundances should be the number of females in each age class.

Another important point is that an age-specific fecundity should be estimated as the average over *all* the individuals (or all females) in that age class. For example, assume that you are modeling the females in a bird population, and you want to estimate the fecundity of the one-year-old age class based on the following information.

1. An average of 2.5 chicks fledged and survived to next year per nest (in which the breeding female was one year old),
2. sex ratio at birth is 1:1 (i.e., ratio of females is 0.5), and
3. only 40% of one-year-old females breed.

In this case, a sex ratio of 1:1 among 2.5 chicks/nest means $0.5 \cdot 2.5 = 1.25$ daughters per *nesting* one-year-old female. But fecundity must be expressed as the average over *all* females in this age class. Because only 40% of one-year-old females breed, the average fecundity for this age class is $1.25 \cdot 0.40 = 0.5$ daughters per female.

4.4 The Leslie matrix

Using the equations we studied in Section 4.3, we can predict the abundance in each age class from time step t to $t+1$. Below, we consider the specific case of predicting abundances for four age classes.

$$N_0(t+1) = F_0(t) \cdot N_0(t) + F_1(t) \cdot N_1(t) + F_2(t) \cdot N_2(t) + F_3(t) \cdot N_3(t)$$
$$N_1(t+1) = N_0(t) \cdot S_0$$
$$N_2(t+1) = N_1(t) \cdot S_1$$
$$N_3(t+1) = N_2(t) \cdot S_2$$

An operation that is equivalent to this set of equations is matrix multiplication. A matrix is a table of numbers arranged in rows and columns. The matrices we will consider in this book are square tables, i.e., the number of rows equals the number of columns. If the parameters (survival rates and fecundities) are arranged in the form of a matrix as shown below, multiplying this matrix with the vector of age distribution at time t gives the age distribution at time $t+1$:

$$\begin{bmatrix} N_0(t+1) \\ N_1(t+1) \\ N_2(t+1) \\ N_3(t+1) \end{bmatrix} = \begin{bmatrix} F_0 & F_1 & F_2 & F_3 \\ S_0 & 0 & 0 & 0 \\ 0 & S_1 & 0 & 0 \\ 0 & 0 & S_2 & 0 \end{bmatrix} \cdot \begin{bmatrix} N_0(t) \\ N_1(t) \\ N_2(t) \\ N_3(t) \end{bmatrix}$$

If you know matrix multiplication, you can confirm for yourself that this multiplication is equivalent to the four equations listed above (and you can skip this paragraph and the next one). If you don't know matrix multiplication, it is easy to learn for this specific case (a square matrix multiplied by a vector). Let's call the vector on the left the result vector (since it is the result of the multiplication). The result vector gives the abundance in all age classes in the next time step ($t+1$). To calculate the *first* element of the result vector [which is, in this case, $N_0(t+1)$, the abundance of zero-year olds in the next time step], you do two things: First you do an element-by-element multiplication of the *first row* of the matrix by the vector on the right. Element-by-element multiplication means that the first number of the row, F_0, is multiplied by the first number of the vector, $N_0(t)$, the second number of the row with the second number of the vector, and so on. Second, you add up all these products:

$$F_0(t) \cdot N_0(t) + F_1(t) \cdot N_1(t) + F_2(t) \cdot N_2(t) + F_3(t) \cdot N_3(t)$$

This operation is the first equation for the population projection we listed above, and gives the first element of the result vector. Next you repeat the same process for other rows. For example, to calculate the second element of the result vector, you do an element-by-element multiplication and summation of the *second* row of the matrix and the vector on the right. However, since there are three zeros in the second row, there is only one nonzero term in the summation (namely $N_0(t) \cdot S_0$). The same goes for the rest of the rows.

As mentioned above, this matrix multiplication is equivalent to the four equations listed at the beginning of this section. You might think that the reason we do this operation with matrices and vectors instead of the set of equations is to make it more difficult. Actually, the reason is that the matrix form makes it easier to expand this model into different types of structures that we will discuss in the next chapter. The matrix that we used in this operation is called a Leslie matrix (named after P.H. Leslie, a population biologist who studied age-structured models in the 1940s) and is represented by

$$L = \begin{bmatrix} F_0 & F_1 & F_2 & F_3 \\ S_0 & 0 & 0 & 0 \\ 0 & S_1 & 0 & 0 \\ 0 & 0 & S_2 & 0 \end{bmatrix}$$

where F_x and S_x are, respectively, the fecundity and the survival rate of the x-year olds, as we discussed in the previous section. The Leslie matrix has a very specific structure. The elements of the top row are fecundities. The survival rates are in the subdiagonal of the matrix. All other elements of the matrix are zeros. Subdiagonal means below the diagonal. The diagonal is the set of four numbers (F_0 and three zeros) that go from the upper-left to the lower-right corner. Often $F_0 = 0$, as in the Helmeted Honeyeater example. The operation of matrix multiplication can be expressed in matrix notation as

$$N(t+1) = L \cdot N(t)$$

where L is the Leslie matrix. Note that when you do this multiplication, the matrix must be on the left and the vector on the right, $L \cdot N(t)$. (The multiplication $N(t) \cdot L$ would be incorrect.)

4.4.1 Leslie Matrix for Helmeted Honeyeaters

We already did the hard part of developing a matrix model for the Helmeted Honeyeaters, by estimating the age-specific parameters. The rest is simply a matter of arranging the survival rates and fecundities in the correct order, which is

$$\begin{bmatrix} 0 & 0.48 & 0.48 & 0.48 & 0.48 & 0.48 & 0.48 & 0.48 & 0.48 & 0.48 \\ 0.703 & 0 & 0 & 0 & 0 & 0 & 0 & 0 & 0 & 0 \\ 0 & 0.717 & 0 & 0 & 0 & 0 & 0 & 0 & 0 & 0 \\ 0 & 0 & 0.751 & 0 & 0 & 0 & 0 & 0 & 0 & 0 \\ 0 & 0 & 0 & 0.769 & 0 & 0 & 0 & 0 & 0 & 0 \\ 0 & 0 & 0 & 0 & 0.746 & 0 & 0 & 0 & 0 & 0 \\ 0 & 0 & 0 & 0 & 0 & 0.717 & 0 & 0 & 0 & 0 \\ 0 & 0 & 0 & 0 & 0 & 0 & 0.806 & 0 & 0 & 0 \\ 0 & 0 & 0 & 0 & 0 & 0 & 0 & 0.778 & 0 & 0 \\ 0 & 0 & 0 & 0 & 0 & 0 & 0 & 0 & 0.667 & 0 \end{bmatrix}$$

Note that this matrix model includes 10 age classes, for ages 0 through 9. In fact, Helmeted Honeyeaters are known to live longer than 10 years. The fact that we did not encounter a 10-year-old bird in our census may be because of the small size of the population. Even if a species lives for many decades, there will be fewer individuals in the older age classes, since some will die every year. As a result, we may not be able to observe very old individuals in a small population. We can correct this by adding a new element to the matrix. The lower-right corner element of the above matrix is zero. If instead it were, say, 0.667 (the same number as S_8), this would mean that 66.7% of the individuals in the tenth age class ("9-year olds") would remain in the tenth age class in the next year. Of course, they wouldn't be nine years old anymore; the tenth age class would consist of individuals aged nine years or older. When individuals of a certain age or older are lumped into one age class, that age class is called a *composite age class*. This is an efficient way to model populations of organisms with indeterminate lifespans. It may be useful for modeling species in which the vital rates do not change much after a certain age, or when the available data do not allow estimation of survival and fecundity rates separately for each age class after a certain age.

Before we rewrite our matrix model of the Helmeted Honeyeaters using a composite age class, there is another improvement we might consider. If you look back at the counts of older age classes in Table 4.1, you will notice that there are few individuals aged 3 and older. This presents a problem in estimating survival rates. Since individuals come only in discrete units, few individuals mean that there may be a lot of sampling error in our estimations of survival rates. One way to get around this problem is to lump all these age classes into one. If we define the composite age class as ages 3 and older (instead of ages 9 and older), we will have a more reliable estimate of the survival rates. From the data we have, we don't have convincing evidence that the survivals are different for older birds, so it is probably reasonable to combine them into a single age class.

Survival rates for composite classes may be calculated by pooling the counts for the appropriate age classes. If we want to pool age class 3 and older, the data in Table 4.1 may be simplified as follows.

Age	1991	1992	1993	1994
0	26	28	27	29
1	16	17	20	20
2	12	11	13	14
3	9	8	9	10
4+	22	19	19	24

Note that, even though we want to pool all individuals age 3 and older into a single age class, we still need the age 3 abundance separately. This is because the number of animals aged 3 and older in 1991 determine the number that are 4 and older in 1992, the number that are 3 and older in 1992 determine the number that are 4 and older in 1993, and so on.

The survival rates of the 3+ age class from 1991 to 1992 is given by

$$S_{3+}(1991) = \frac{N_{4+}(1992)}{N_3(1991) + N_{4+}(1991)} = \frac{19}{(22+9)} = 0.613$$

This represents the proportion of individuals 3 years old and older in 1991 that survived until the next census in 1992. Similar calculations give survival rate of this composite age class from 1992 to 1993, and from 1993 to 1994:

$$S_{3+}(1992) = \frac{19}{(19+8)} = 0.704 \qquad S_{3+}(1993) = \frac{24}{(19+9)} = 0.857$$

The average of these three numbers is 0.725. (You may also calculate a weighted average; see Section 4.7.1.1 under Additional Topics.) Our new model, then, has only four age classes: 0, 1, 2, and 3+. Our new matrix is:

	Age 0	Age 1	Age 2	Age 3+
Age 0	0	0.48	0.48	0.48
Age 1	0.703	0	0	0
Age 2	0	0.717	0	0
Age 3+	0	0	0.751	0.725

The element in the lower-right corner of the matrix is S_{3+}, the average survival rate of three-year-old and older individuals that we just calculated.

4.4.2 Projection with the Leslie Matrix

We will now use the matrix we found to predict the age structure of the population (i.e., the abundances in each age class). After combining the counts for ages 3 and above into a single age class, the vector of age distribution for 1994 (from Table 4.1) becomes 29, 20, 14, 34. To project this population, we multiply this vector with the above matrix:

$$\begin{bmatrix} 0 & 0.48 & 0.48 & 0.48 \\ 0.703 & 0 & 0 & 0 \\ 0 & 0.717 & 0 & 0 \\ 0 & 0 & 0.751 & 0.725 \end{bmatrix} \cdot \begin{bmatrix} 29 \\ 20 \\ 14 \\ 34 \end{bmatrix} = \begin{bmatrix} 33 \\ 20 \\ 14 \\ 35 \end{bmatrix}$$

Note that at every step of this matrix multiplication, we round the result to the nearest integer. For example, the calculation for the zero-year olds is

$$0.48 \times 20 + 0.48 \times 14 + 0.48 \times 34 = 9.60 + 6.72 + 16.32$$
$$= 10 + 7 + 16$$
$$= 33$$

The total number of individuals predicted to be in the population in the next time step is 102. Since the previous total was 97, this gives a growth rate of $102/97 = 1.052$, or 5.2% growth in one year.

An interesting and important characteristic of age-structured dynamics is that the growth of the population depends on the initial age distribution (i.e., the distribution of individuals among age classes at the initial time step). We will demonstrate this with two examples, in which we predict the age structure of the population starting with the same total number of individuals (97), but with two different initial age distributions. The first age distribution has about an equal number of individuals (24 to 25) in each age class. The projection,

$$\begin{bmatrix} 0 & 0.48 & 0.48 & 0.48 \\ 0.703 & 0 & 0 & 0 \\ 0 & 0.717 & 0 & 0 \\ 0 & 0 & 0.751 & 0.725 \end{bmatrix} \cdot \begin{bmatrix} 24 \\ 24 \\ 24 \\ 25 \end{bmatrix} = \begin{bmatrix} 36 \\ 17 \\ 17 \\ 36 \end{bmatrix}$$

predicts a total of 106 individuals (36 + 17 + 17 + 36) in the next year, which gives a one-year growth rate of $106/97 = 1.093$, or a 9.3% increase. The second example has a much more unequal initial distribution of individuals among age classes, with only 10 individuals in each of the three older classes, and the rest (67 individuals) in zero-year-old age class. This projection,

$$\begin{bmatrix} 0 & 0.48 & 0.48 & 0.48 \\ 0.703 & 0 & 0 & 0 \\ 0 & 0.717 & 0 & 0 \\ 0 & 0 & 0.751 & 0.725 \end{bmatrix} \cdot \begin{bmatrix} 67 \\ 10 \\ 10 \\ 10 \end{bmatrix} = \begin{bmatrix} 15 \\ 47 \\ 7 \\ 15 \end{bmatrix}$$

predicts a total of 84 individuals in the next year. This gives a one-year growth rate of $84/97 = 0.866$, which is a 13.4% decline in one year. The same matrix (i.e., the same set of survival rates and fecundities) in equal-sized populations predicted both a substantial growth and a substantial decline in the population size, depending on how individuals were distributed among age classes in the population. The next section explains the cause of this apparent anomaly.

4.4.3 Stable Age Distribution

If we continued the last projection above (with the unequal initial distribution of individuals among age classes), we would multiply the same matrix with the next year's age distribution vector (15, 47, 7, 15). This would give the age distribution for a third time step, which we would multiply again with the same matrix, and so on. We can plot the result as a trajectory for each age class, as in Figure 4.3.

This projection was made with the same Leslie matrix we used above, in which we ignored all forms of stochasticity. The fluctuations in the abundance of various age classes do not result from variation in matrix elements (survivals and fecundities), but from the particular distribution of individuals among age classes. Note that the fluctuations subside after the fifth year, and all age classes start to grow more-or-less in parallel. We can look at this in another way, by plotting the *proportion* of individuals in each age class (Figure 4.4). In this figure the total is always 1.0, and the areas show the relative abundances of individuals in different age classes. The projection started with about 10% (10 out of 97) in each of the three older classes (one-year, two-year, and three-plus-year olds). After the fifth year, the proportion in each age class becomes stabilized (notice that year 7 and year 50 have the same distribution), even though the population is growing and the abundance in each age class keeps changing, as we observed in the previous figure.

Repeatedly multiplying an age distribution by a Leslie matrix with constant elements tends to draw it to a special configuration known as the *stable age distribution*. Before it reaches the stable age distribution, the population may show considerable fluctuations. These are especially pronounced if reproduction is concentrated in one or two older age classes. Note that these fluctuations are not caused by changes in the environment, but result from the distribution of individuals within the population. Of course, the reason that the population is not at the stable distribution may have something to do with the environment. For example, if there is a sudden influx of individuals, the age structure may be changed. It may also be changed if the fecundities suddenly increase, as happened in the human population in the United States and elsewhere around the world after World War II, generating the population fluctuation known as the "baby boom."

The age structure may also change gradually, in response to trends in vital rates. For example, as people live longer (i.e., the survival rates of older age classes increase), the proportion of the population in older age classes increases. Such changes have important consequences, e.g., for social welfare programs for older people.

120   Chapter 4 Age Structure

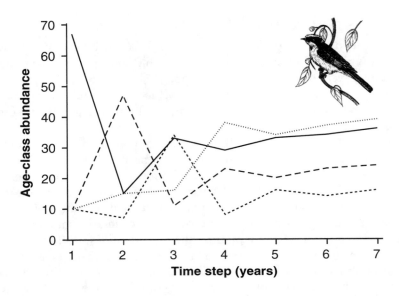

Figure 4.3.   The predicted abundance of zero-year-old (solid curve), one-year old (long dashes), two-years-old (dashes), and three-years-old and older (dots) Helmeted Honeyeaters.

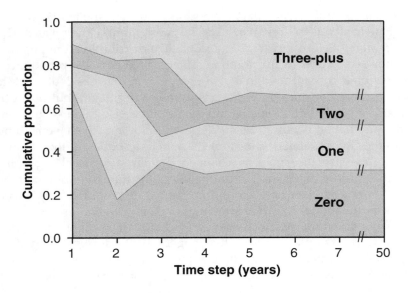

Figure 4.4.   The proportion of zero-year-old, one-year-old, two-years-old, and three-years-old and older Helmeted Honeyeaters.

Once a population reaches the stable age distribution, the proportion of individuals in each of the age classes remains the same. If the population is growing or declining, all age classes (as well as the total population size) grow or decline at the same rate. At this point, multiplying the age distribution by the Leslie matrix is the same as multiplying it by a scalar number. Mathematically, this is

$$N(t+1) = L \cdot N(t) = \lambda \cdot N(t)$$

where λ (lambda) is a special number called the dominant eigenvalue of the matrix. Once at its stable age distribution, the population grows exponentially with rate λ. The stable age distribution is stable because, if the population is perturbed from this configuration, it will spontaneously return to it over time, if the matrix elements remain constant. However, if a population is cycling because of density dependence, or if the matrix elements are fluctuating because of environmental factors, the age distribution may not tend toward a stable configuration over time.

The dominant eigenvalue λ measures the asymptotic or deterministic growth rate of the population, which tells how the population would be changing if the parameters in the model were constant for an indefinite length of time. λ is often called the finite rate of increase; it is equivalent to the constant growth rate R for deterministic exponential population growth we used in Chapter 1. As we demonstrated with the above examples, population growth at any particular time step is not always given by the dominant eigenvalue, even if the matrix elements stay the same. In fact, it is only when the population is at its stable distribution that the population's overall growth is measured by λ. When the initial distribution is different from the stable form, the abundances at the next time step must be computed by working out the matrix multiplication of the Leslie matrix by the current age distribution.

4.4.4 Reproductive Value

Both the dominant eigenvalue λ and the stable age distribution are properties of a Leslie matrix; in other words, they are determined by the matrix itself and do not depend on the abundances, or any other parameter. Another variable that we can calculate based on a Leslie matrix is reproductive value (Fisher 1930), which is an age-specific measure of the relative contribution of each age class to future generations. It is the number of offspring an individual in a given age class will produce, including all its descendants. The reproductive value is expressed as relative to the reproductive value of an individual in the first age class. Thus reproductive value for the first age class is always 1.0. The calculation of reproductive value is quite tedious, so we will not cover its formula. The program RAMAS EcoLab allows the calculation of reproductive value (in addition to stable age distribution, and the finite rate of increase, λ) of a matrix model; we will demonstrate these in the exercises.

Reproductive values depend on both survival rates and fecundities. You might think that younger classes should have a higher reproductive value since they have a longer life (and a longer reproductive life) than older individuals. But very young individuals often do not reproduce, and may die before they begin reproducing. Or, you might think that the age class with the highest fecundity should have the highest reproductive value. Often older age classes have the highest fecundities, but they may not have as long a reproductive life left as younger individuals.

Knowing the relative reproductive values of different age classes may be important in several practical cases, such as harvesting. Often harvesting may have a smaller long-term effect on the population when only the age classes with lowest reproductive values are harvested.

Another case where reproductive values may be useful is the reintroduction of individuals to a location where the species has become extinct. It may be more efficient to reintroduce individuals from age classes with high reproductive value, instead of younger individuals who may die before they reach reproductive age, or older individuals who may not have a long reproductive life left. However, reproductive value cannot be the only consideration in such a decision. Also important are spatial considerations (the specific locations where such introduction takes place) and the effects of age distribution on population fluctuations. We will discuss the spatial considerations in a later chapter. In Section 4.4.3, we demonstrated how abundances can fluctuate (even in the absence of any environmental effects) when the distribution of individuals among age classes differs from the stable age distribution. If a population is started with all individuals in the same age class (for example the age class with the maximum reproductive value), it will be far away from the stable age distribution, and may fluctuate quite a bit before settling into the stable age distribution. Such fluctuations may carry the population close to dangerously low levels, or may cause uneven and rapid depletion of its resources.

Another consideration might involve the cost of the reintroduction to the source population. If individuals with the highest reproductive value are taken out of a source population for reintroduction elsewhere, the risk of decline of the source population might increase. Of course, reintroduction helps the target population. Whether this balances the cost to the source population can only be analyzed with a model that includes both populations.

Practical questions such as these rarely have formulaic answers that apply to all cases. Rather, they often require case-specific analyses. Modeling the effects of each management option (for example, different age distributions of introduced individuals), and comparing the model results in terms of the potential for increase or persistence, provides case-specific answers to these questions.

4.5 Adding Stochasticity

So far our matrix models do not incorporate any of the various types of uncertainties we discussed in Chapter 2. In this section, we will explore ways to incorporate some of these uncertainties into Leslie matrix models.

4.5.1 Demographic Stochasticity

When the number of individuals gets to be very small, there is a source of variation that becomes important even if the vital rates remain constant. This is exactly the same sort of variation we discussed in Chapter 2 when we added demographic stochasticity to the Muskox model. Here, we will apply the same methods to age-structured models. Suppose that the survival rate of 3-year olds in a particular population is 0.4. If there are one hundred 3-year olds, then the number of 4-year olds next year will be about 40. However, if there are only three individuals in the age class, the number of 4-year olds next year will not be 1.2, because you cannot have a fraction of an individual. We took care of this fact to some extent in our Helmeted Honeyeater model above, by rounding the number of individuals at each step of the calculation. However, just rounding to the nearest integer is not enough, as we will see.

One way of interpreting a survival rate of 0.4 is to say that each individual has a 40% chance of surviving. But each individual can either live or die. We can guess the number that will survive by following the fate of each individual, as we did in Chapter 2 for the Muskox population. In the above example, if there are three individuals, each with a 40% chance of surviving, we can decide on the fate of each individual by selecting a uniform random number between 0 and 1 and checking to see if it is greater or less than 0.4, as we did in Chapter 2. If we repeated this experiment several times, we could end up with 1 survivor out of the three individuals at one time, 2 survivors at other times, and even 0 or 3 survivors once in a while. The distribution of the number of survivors after many such repeated trials is shown in Figure 4.5. The mean of this distribution is 1.2, which is what you would expect if you ignored demographic stochasticity. However, each value of this distribution is either 0, 1, 2, or 3. Such statistical distributions that give only integer values are called discrete distributions.

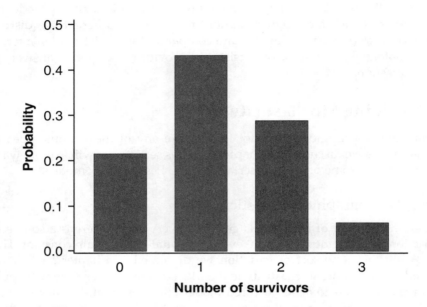

Figure 4.5. Binomial distribution showing the probability of 0, 1, 2 and 3 survivors with an initial population of 3 individuals and a survival rate of 0.4.

This type of distribution, the binomial distribution, applies to cases where there are two possible outcomes (survival or death, for example). You can calculate the probability that all will survive by multiplying $0.4 \cdot 0.4 \cdot 0.4 = 0.4^3$, and you can also easily calculate the probability that all of them will die: $(1-0.4)^3$. We can make these multiplications when we assume that the fates of these individuals are independent, given that the environment stays constant. When three events (individual 1 survives, individual 2 survives, individual 3 survives) are independent of each other, their joint probability (the probability that all three will survive) is the product of their individual probabilities. We couldn't make this assumption of independence if, for example, we knew that two of the individuals were dependent offspring of the third one.

We could calculate the probability of the other two outcomes (1 out of 3 and 2 out of 3 surviving) in a similar, but slightly more complicated way. But, we will leave this tedious computations for the computer to do.

Notice that if the number of individuals is high (for example 100), extreme events (all survive, or all die) will be very unlikely, because their probability would be equal to 0.4 (for all surviving) or 0.6 (for all dying) multiplied by itself 100 times. Either of these is a very small number. In general, demographic stochasticity is less important (compared to other forms of stochasticity) when the population is large.

Demographic stochasticity also applies to fecundities. Mothers cannot have a fraction of an offspring; they can only have a whole number of offspring. This can be modeled in the same way we did before, if each mother had either one offspring or none at all (i.e., if there are only two possible outcomes, we could use the binomial distribution). If the largest possible number is not 1, we cannot use the binomial distribution. Instead, we use another discrete distribution, called the Poisson distribution (which we do in RAMAS EcoLab). Learning how to use Poisson distribution without the help of a computer is beyond the scope of this book.

We will demonstrate the effect of including or excluding demographic stochasticity in the exercises at the end of the chapter. In general, demographic stochasticity should be included in all models unless the model describes densities (such as number of animals per km^2) instead of absolute numbers of individuals.

4.5.2 Environmental Stochasticity

Natural environments often change in an unpredictable fashion, causing changes in a population's demographic characteristics, such as survivals and fecundities. If we knew which environmental factors affected which population parameters, *and* we knew how much they affected these parameters, *and* we knew how these environmental factors would change in the future, then we could explicitly incorporate these factors into our prediction of the population's future. Such detailed knowledge of biology, meteorology, and their interaction is clearly impossible at the present time, and even in the foreseeable future.

The crude approximation to modeling the effects of environmental fluctuations in computer-implemented models involves replacing the constant parameters, such as survival rates and fecundities, with random variables. We cannot know what the exact parameters will be from year to year, but we can estimate from past observations what their average values will be and the ranges over which they might vary. We can use the mean and variance of the parameters to help us predict population abundance in the future.

This is the same approach that we used in Chapter 2. But in the case of an age-structured model we have several parameters, including several age-specific survivals and fecundities. Each of them can vary over time in response to the environment. In the Helmeted Honeyeater example, we had three estimates for each survival rate. We used these three numbers to estimate an average of the survival rate. We can also use the same set of numbers to estimate their standard deviation. The standard deviation of survival rates and fecundities are given in Table 4.3; these numbers are based on data in Table 4.1 and on the assumption that the fecundities of all breeding age classes (one and above) are the same. Remember from Chapter 2 that estimates of variation in a model often assume that all observed variation is due to the environment. We have made that assumption here. While the Helmeted Honeyeater population may be sufficiently large and well known that sampling error and demographic stochasticity are negligible, these sources of variation usually are present and ideally should be removed from the estimates of environmental variation.

Table 4.3. Standard deviation of age-specific vital rates in the Helmeted Honeyeater model.

Age (x)	Standard deviation of survival rate	Standard deviation of fecundity
0	0.0364	0.0
1	0.0338	0.0075
2	0.0631	0.0075
3+	0.1233	0.0075

How do we use these standard deviations in a matrix model? At every time step, before making the matrix multiplication we discussed above, we sample the elements of the matrix (survival rates and fecundities) from random distributions. We specify this random selection process such that in the long run the sampled survival rate of, for example, zero-year olds will have an average of 0.703 and a standard deviation of 0.0364. Because the sampled values change at each time step, the population growth will show some variation (as we demonstrated in Chapter 2). Of course, since the survival rates and fecundities are chosen at random, we would have little confidence that any one simulated trajectory would actually occur. This is because the trajectory would probably be different if we did the simulation again. To get a prediction out of these simulations, we need to repeat them many times. Then, even though we do not trust any particular trajectory to represent the future closely, we could argue that the set of many trajectories describe some statistical features of the population's future behavior. For instance, we can estimate a mean trend in abundance. And we can predict the magnitude of year-to-year fluctuations that the population may exhibit even if we cannot say confidently which years will have highs and which will have lows.

An important point about how these random survival rates and fecundities are selected is whether they are correlated or not. A positive correlation between survival rates means that if there is a low survival rate for zero-year olds in a particular year, it is likely that there will also be a low survival rate for other age classes. We will talk about correlations in a later chapter in another context, but for now we will just make a simplifying assumption. In RAMAS EcoLab, we assume that all vital rates (survivals and fecundities) are perfectly correlated. In other words, a "bad" year means that all survivals and fecundities are lower than their respective averages, and a "good" year means they are all higher than average. It is possible to make other assumptions, or to specify how exactly they should be correlated, but this complicates the models considerably.

4.6 Life Tables

Suppose we identified 1,000 newborn individuals, followed them through their lifetimes, until all of them died, and at each time step (for example, each year) recorded the number of these individuals that were still alive, and the number of offspring they produced. This is possible to do if organisms can be individually identified, if there is no emigration, if the parents of all offspring can be identified, and there is no immigration. An example would be to sow 1,000 seeds of a perennial plant in a plot where we are reasonably sure that there are no other seeds to start with. We would then tag each seedling that comes up the following year, and every year we would count the number of plants, and the number of seeds they produce. We would assume that all plants develop seeds at the same time, and that we census the population every year immediately after seed production. We also need to be sure that all seeds germinate or die before the second census (i.e., there is no "seed bank"). Such a data set would look like Table 4.4.

This is called a *cohort life table*. A cohort is a group of individuals born at the same time (or within a short interval of time, for example in the same breeding season). A cohort life table describes the demography of a single cohort. It is also called a *dynamic life table*, since it follows individuals through time. In the table, the first column shows the age (in years, denoted by the symbol x), and the second column shows the number alive at the beginning of that age (N_x). The first age is age zero, and the number alive is 1,000, since this is the starting number of individuals. The third column shows the number of offspring they produced at a given age (B_x). These are all the data; the rest of the table shows various variables calculated from these data.

Table 4.4. Life table for a plant species.

(1)	(2)	(3)	(4)	(5)	(6)	(7)	(8)	(9)
x	N_x	B_x	l_x	S_x	m_x	$l_x \cdot m_x$	$x \cdot l_x \cdot m_x$	$e^{-rx} \cdot l_x \cdot m_x$
0	1000	0	1.000	0.186	0.00	0.000	0.000	0
1	186	0	0.186	0.312	0.00	0.000	0.000	0
2	58	690	0.058	0.586	11.90	0.690	1.380	0.4730
3	34	465	0.034	0.647	13.68	0.465	1.395	0.2639
4	22	314	0.022	0.545	14.27	0.314	1.256	0.1475
5	12	201	0.012	0.417	16.75	0.201	1.005	0.0782
6	5	87	0.005	0.400	17.40	0.087	0.522	0.0280
7	2	35	0.002	0.000	17.50	0.035	0.245	0.0093
8	0	0	0.000	0	0	0	0	0

$$\text{Total} = R_0 = 1.792 \quad 5.803 \quad 1.0000$$
$$T = 3.238$$
$$r_{est} = 0.180$$
$$r = 0.189$$
$$R = 1.208$$

4.6.1 The Survivorship Schedule

The fourth column of the table (labeled l_x) gives the survivorship schedule, which is the proportion of the original number of individuals in the cohort that are still alive at the beginning of age x. Survivorship for age 0 is by definition equal to one. The survivorship to age x is calculated by dividing the number alive at age x (N_x; the second column) by the starting number of individuals (1000 in this case):

$$l_x = \frac{N_x}{N_0}$$

Be careful not to confuse survivorship (l_x) with the survival rate (S_x; column 5 in the table) that we used in the Leslie matrix, although both are age-specific rates, and both can be expressed as probabilities. Survival rate is the probability of surviving *from a given age to the next*, whereas survivorship is the probability of surviving *from birth to a given age*. For example, survivorship to age 2 is calculated as

$$l_2 = \frac{N_2}{N_0} = \frac{58}{1000} = 0.058$$

whereas the survival rate for age 2 (from age 2 to age 3) is calculated as

$$S_2 = \frac{N_3}{N_2} = \frac{34}{58} = 0.586$$

To calculate survival rate for age x from survivorship, divide the survivorship for age $x+1$ with the survivorship for age x:

$$S_x = \frac{l_{x+1}}{l_x}$$

To calculate survivorship (to age x) from survival rates, you need to multiply all survival rates up to, but excluding, x:

$$l_x = S_0 \, S_1 \, \cdots \, S_{x-1}$$

This is because the survivorship is the probability of surviving from birth to the beginning of age x, for which the individual must have survived from age 0 (birth) to age 1 (S_0), then from age 1 to age 2 (S_1), etc., and finally, from age $x-1$ to age x (S_{x-1}).

The survivorship schedule is a monotonically decreasing function, which means that as you increase x, l_x either decreases or stays the same, but does not increase. (Can you explain why?). The plot of l_x as a function of x is called a survivorship curve; its shape characterizes the life history of a species. For example, consider the curves in Figure 4.6 (note that the survivorships are in logarithmic scale).

The shape of the survivorship curve is a function of the distribution of mortality among age classes. If mortality is quite low for most of a species' life, and gets high only at the end, the result is a Type I survivorship curve, which is typical of human populations. In some species, the mortality is much higher in younger age classes, and lower in older classes, giving a Type III curve. If mortality is constant throughout a species' lifetime (i.e., all age classes have approximately the same survival rate), the result is a Type II curve. All three of these curves are simplifications. In reality, most species have survivorship curves that are intermediate between, or a mixture of, two or three of these types. For example, the survivorship curve for *Orchesella cincta*, a forest insect (the dotted curve; van Straalen 1985) indicates that this species has a relatively high mortality in the youngest age class (the curve starts with a steep decline, as in the Type III curve), relatively low mortality

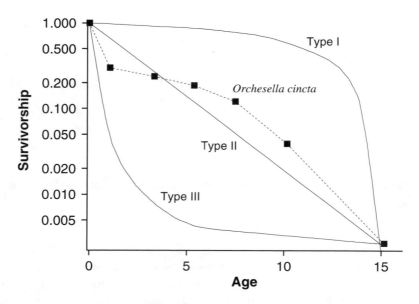

Figure 4.6. Three idealized types of survivorship (l_x) curves as a function of age, x (solid curves), and survivorship curve for *Orchesella cincta*, a forest insect, with age in weeks (dotted curve; data from van Straalen 1985).

in middle ages (the curve levels off, as in the Type I curve), and relatively high but constant mortality in older ages (the curve ends as in a Type II curve).

4.6.2 The Maternity (Fertility) Schedule

The third column in Table 4.4 gives the total number of seeds produced by plants in each age class. To calculate fertility, we divide these numbers with the corresponding number of individual plants in each age class. The results are in column 6, labeled m_x, and gives the average number of seeds produced by a plant of age x.

It is important to note the difference between fertility (m_x) and fecundity (F_x) that we use in a Leslie matrix. Fertility gives the number of offspring (e.g., seeds) produced by an individual in a given breeding season. The fecundity, F_x, is the average number, per individual of age x alive at a given time step, of offspring censused at the next time step. Fecundity values incorporate two kinds of mortality over the time step. Some of the individuals (plants) that were alive during the last census die before reproducing, and some of the offspring (seeds) that are produced die before they can be counted in the next census. The difference will become clearer when we consider how to construct a Leslie matrix from life table data, later in this chapter.

4.6.3 Life History Parameters

There are a number of other life history parameters that can be calculated based on the survivorship schedule l_x and the fertility schedule m_x. One of these is the replacement rate, or net reproductive rate (R_0). Do not confuse this with the *replacement curve* (see Chapter 3), or with the growth rate (R). Net reproductive rate is a measure of the expected number of offspring produced by an individual over its lifetime (for a female-only model, it is a measure of the expected number of daughters produced by a female over her lifetime).

The net reproductive rate is calculated by summing up the product $l_x \cdot m_x$ (which is given in column 7 of Table 4.4) over all age classes:

$$R_0 = \sum l_x m_x$$

In this and the following equations, the sigma symbol (Σ) indicates summation over all age classes. Another useful life history parameter is the generation time (T_G), which is a measure of the average age of reproduction. To calculate generation time, first calculate the product $x \cdot l_x \cdot m_x$ (which is given in column 8 of the table) for each age class, then add these products over all age classes. Finally divide this number by the net reproductive rate.

$$T_G = \frac{\sum x l_x m_x}{\sum l_x m_x} = \frac{\sum x l_x m_x}{R_0}$$

Generation time and net reproductive rate allow the computation of the finite rate of increase that we introduced in Chapter 1, and discussed above for the Leslie matrix. In life table calculations, it is usually another measure of growth rate that is calculated. This is the "instantaneous rate of growth" (r), which is approximately related to the finite rate of increase as

$$r = \ln(R)$$

or

$$R = e^r$$

The instantaneous rate of growth (*r*) is calculated by finding the value of *r* that satisfies the following equation (assuming the index for the youngest age class is 0):

$$\sum e^{-rx} l_x m_x = 1$$

To find *r*, you start with an estimate r_{est}, and calculate the above equation. If the result is greater than 1.0, you *increase* r_{est}; if it is less than 1.0, you decrease r_{est}. Then you evaluate the equation again, until the sum is 1.0 (this iterative process may take a long while with a calculator, but it is quite easy to do if you know how to use spreadsheet software). A good initial estimate for r_{est} is

$$r_{est} = \frac{\ln(R_0)}{T_G}$$

Our table of the plant population shows the calculation of these life history parameters under column 7 and in column 9.

4.6.4 Life Table Assumptions

A cohort life table such as the one we have been analyzing makes a very important assumption. It assumes that as a cohort ages, the vital rates change only as a function of age. In other words, the observed difference in survivorship and fertility in different time steps is because the individuals are aging. We know from Chapter 2 that they may also be changing because of changes in the environment. For example, according to the table, the survival rate was 0.586 in year 3 (when the plants were 2 years old) and 0.647 in year 4 (when the plants were 3 years old). We interpreted this difference as 2-year olds having a lower survival rate than 3-year olds. Another possibility is that year 3 was a worse year for survival of these plants than year 4. There is no way of knowing which explanation is true, unless this was an experiment in controlled laboratory conditions.

Compare this with the type of data we used for constructing a Leslie matrix for the Helmeted Honeyeaters earlier in the chapter. Because all age classes were present in all years, we were able to estimate the survival rates for multiple years and get an idea of their fluctuations due to the changes in the environment. In a cohort life table, one age class is present only in one year, making it difficult (if not impossible) to separate the effects of age versus environment on the vital rates. Because we do not know the effect of the environment on survivorship and fertility schedules in a cohort life table, we cannot trust the various life history statistics (net reproductive rate, instantaneous rate of increase, etc.) calculated from these schedules. They will give a summary of what happened during the lifetime of a particular cohort, but they may not be representative of the dynamics of the population in general.

If this is the case, you may wonder why we spent all this time discussing life tables. Unfortunately, data are often in short supply in ecology. Often, a cohort life table may be the only demographic data available for a particular species. In such a case, it is important to be able to make as much use of such data as possible, without forgetting the assumptions of models we construct based on such data. In the next section, we will discuss how life table data can be used to construct a Leslie matrix.

In the preceding sections, we concentrated on cohort (or, dynamic) life tables. Another type of table is called a "static life table," and consists of counts of individuals in different age classes at one time step (i.e., like a snap-shot of the population). To imagine this, replace the N_x column of the cohort life table with *one of the columns* of Table 4.1 (a Helmeted Honeyeater census at one particular year). If you calculated the life table statistics of such a table several times, each time with another year's census, you would get different results. There are two reasons for this. First, the survival rates are changing every year (as we calculated) as a result of environmental fluctuations. Second, even if the environment did not change, the proportion of individuals in each age class change because the population is not at its stable age distribution. So, two conditions are necessary for a static life table to give results representative of the long term future of the population: the vital rates (i.e., environment) must stay relatively constant from one year to the next, and the population must be at its stable age distribution. These two conditions are rarely met in nature, so static life tables are even less reliable than cohort life tables.

4.7 Additional topics

4.7.1 Estimating Survivals and Fecundities

In this chapter, we discussed simple methods of estimating survival rates and fecundities. In this section, we mention a few more advanced methods for estimating these vital rates from data.

4.7.1.1 Weighted Average for Survival Rates

In Section 4.3.1 we used a simple arithmetic average of the three consecutive estimates of S_0 (the survival rate of zero-year olds). The simple arithmetic average is the sum of three ratios, divided by three:

$$S_0 = \frac{1}{3} \cdot \left(\frac{N_1(1992)}{N_0(1991)} + \frac{N_1(1993)}{N_0(1992)} + \frac{N_1(1994)}{N_0(1993)} \right)$$

This was justified because all three estimates were based on similar numbers of individuals (26 to 28). However, if the number of individuals in the denominators are very different, it is better to use a weighted average. This is because, in general we want averages to be influenced more by estimates based on larger sample sizes. To calculate a weighted average, we simply multiply each number with a weight (W) and divide the sum with the sum of weights:

$$S_0 = \frac{1}{(W_1 + W_2 + W_3)} \cdot \left(W_1 \cdot \frac{N_1(1992)}{N_0(1991)} + W_2 \cdot \frac{N_1(1993)}{N_0(1992)} + W_3 \cdot \frac{N_1(1994)}{N_0(1993)} \right)$$

Note that if all weights are equal to one ($W_1 = W_2 = W_3 = 1$), then this formula is the same as the previous one.

What should the weights (W) be? The simplest option is to make them equal to the denominator, i.e., the number of individuals on which the survival rate is based. If you substitute the appropriate $N_0(t)$ for each W in the above formula, and simplify, you get:

$$S_0 = \frac{N_1(1992) + N_1(1993) + N_1(1994)}{N_0(1991) + N_0(1992) + N_0(1993)}$$

In the Helmeted Honeyeater example in Table 4.1, the average survival of zero-year olds is

$$S_0 = (17 + 20 + 20) / (26 + 28 + 27) = 0.704$$

which is slightly different from the simple average of 0.703. If the differences among the counts in different years were larger, the difference between the simple and weighted averages would also be larger.

4.7.1.2 Mark-recapture

Suppose you caught 100 birds from an isolated population, marked them with bands and released them back to the same population. One year later, you again catch 100 birds, and observe that 10 of them have bands from last year (i.e., they are recaptured). This amount of information does not allow you to estimate a survival rate, because obviously you might not have caught all the marked birds that are still alive. In fact, if you come back the following year, and catch 100 birds, you might recapture some birds that you have marked in year 1 but did not recapture in year 2.

However, if you continue this study for several years in a row, you can use this information to estimate survival rates. The statistical methods used in mark-recapture analyses do this by estimating the probability of recapturing a marked animal, in addition to the probability that it is still alive (i.e., the survival rate). A detailed discussion of mark-recapture analysis requires considerable statistical background and is beyond the scope of this book. However, such analyses are facilitated by specialized software such as CAPTURE, JOLLYAGE, and MARK. You can read about the program MARK at http://www.cnr.colostate.edu/~gwhite/mark/mark.htm. Pollock et al. (1990) provide an extensive review of the topic; for a summary of more recent developments, see Burnham and Anderson (1992), and Lebreton et al. (1993).

4.7.1.3 Estimating Fecundities with Multiple Regression

If we suspected that different age classes might have different fecundities (in other words, if we did not want to assume that $F_1 = F_2 = F_3 = ... = F_9$ as we did in the example in Section 4.3.2), then we would need a way to calculate these different values. We might, for example, do a multiple regression analysis.

Regression is a statistical method for finding a relationship between a dependent variable (in this case, number of offspring surviving to the next census), and an independent variable (in this case, number of potential breeders). Multiple regression is used when there are several independent variables (in this case, number of breeders in each age class).

In the following hypothetical example, the species has just two adult breeding ages, 1 and 2.

Year	N_0	N_1	N_2
0	80	21	14
1	85	11	5
2	45	18	11
3	73	28	18
4	104	15	14
5	90	23	14
6	88	19	9
7	55	17	12
8	52	27	10

Data from field censuses are rearranged as follows

Year	$N_0(t)$	$N_1(t-1)$	$N_2(t-1)$
1	85	21	14
2	45	11	5
3	73	18	11
4	104	28	18
5	90	15	14
6	88	23	14
7	55	19	9
8	52	17	12

The data for breeders (N_1 and N_2) are shifted down one year in relation to the data for recruits (N_0). This is because the number of recruits this year is predicted by the number of breeders in the previous year. This relationship is expressed more formally as a regression model:

$$N_0(t) = b_0 + b_1 N_1(t-1) + b_2 N_2(t-1) + \text{error}$$

A solution for the coefficients, b, is found that best explains the data. The first regression coefficient (b_0) is the constant term, and should be set to zero, unless there is evidence of zero-year-old immigrants from outside the population studied. The coefficients b_1 and b_2 are the age-specific fecundities F_1 and F_2. In this example, the regression analysis gives the relationship:

$$N_0(t) = 0.66\ N_1(t-1) + 5.0\ N_2(t-1)$$

This relationship explains about 78% of the variation in the number of recruits at each census. Any such analysis should be tempered by biological knowledge. For example, this result should concur with direct and indirect field observations that two-year olds are much more successful at reproduction than one-year olds.

Such results are very sensitive to errors in the data. Removing just the last observation changes the estimate of the coefficients to 0.3 and 5.7, respectively. That is, the fecundity estimate for one-year olds is halved. A complete treatment of multiple regression is well outside the scope of this book. For more information, see Sokal and Rohlf (1981).

4.7.2 Estimating a Leslie Matrix from a Life Table

Our focus in this section is using life table data to construct Leslie matrices. As we discussed above, this may be necessary because in some cases you may not have the type of census data we used for the Helmeted Honeyeater. The methods we will discuss below can be useful even if you have such census data. In some cases, younger age classes may be difficult to census because of their size. In other cases, the census method may work only for the breeding population (for example, only territorial owls respond to calls by surveyors; juvenile salmon disperse to the ocean and cannot be censused before they return to rivers to breed). In such cases, fecundities (F_x) may need to be estimated based on measures such as number of chicks fledged per nest, average litter size, belly counts, or a comparable measure that tallies newborns. If you have such measures (which we call *fertilities* or *maternities*, m_x), you must modify these values to use them in a Leslie matrix. How you do this depends on the scheduling of censuses in relation to mortality and reproduction, and on the definition of age of an individual. In other words, when using life table data to construct a Leslie matrix, you need to be aware of the timing of the census in relation to the breeding season. In Figure 4.7, the large black dots represent breeding. Remember that we are assuming a birth-pulse population, in which all breeding takes place in a short period of time. The dotted lines represent reproduction, and the solid lines represent the survival of each cohort.

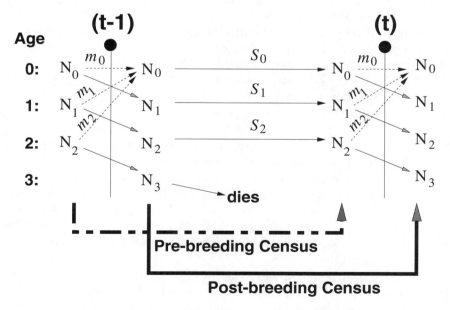

Figure 4.7. The scheduling of census in an age-structured population.

Assume that we are studying an animal species that breeds once a year, and lives for three years; in other words, individuals die after reaching their third birthday but before reaching their fourth birthday. We will define "age" as follows. Individuals within the first year of life are "zero-year olds." These individuals become "one-year olds" immediately after the next breeding season. This is represented by the solid arrow that goes down from N_0 before the census (on the left side of the vertical line) to N_1 after the census (on the right side of the vertical line).

It is important to be clear about how age is defined: In some studies, individuals that have survived their first winter are often called one year old, even if it is before their actual first birthday. Thus newborns become one-year olds *just before* the next breeding season. Here we call such an individual zero-year old *until after* the breeding season. The two definitions of age do not make any difference to the numerical values of the elements of the resulting Leslie matrix, as long as one definition is consistently followed. Keep in mind that as a result of the definition we adopted, m_0 does not refer to the fertility of newborns; it refers to the fertility of individuals that have lived for almost a year.

The definitions of various parameters are the same as earlier in this chapter: The age-specific parameter S_x is the survival rate from age x to age $x+1$, so S_0 is the proportion of newborns that survive to become one-year olds. The age-specific parameter m_x is the maternity (fertility) rate, which is the number of offspring per individual of age x. If only females are modeled, it is number of daughters per mother (see the section *Sex ratio* above). The abundance of x-year olds is represented by N_x, and t denotes the time step.

We will now consider two separate cases. For the first case, assume that we census the population just before the annual reproduction (a "pre-breeding" census); and concentrate on the two columns to the left of the two breeding points at times $t-1$ and t. We assume that there is no mortality between the census and the subsequent breeding. Note that in this census the youngest animals we'll census will be almost (but not quite) 12 months old. According to our definition of age classes, this first age class is zero-year olds, the second age class is one-year olds, etc. At time t, the number of one-year olds will be the number of zero-year olds in the previous census, times the survival rate of one-year olds during the past 12 months: $N_1(t) = N_0(t-1) \cdot S_1$. So, the first survival we use in our Leslie matrix (second row, first column) should be S_1.

Next, we need to compute fecundities for the Leslie matrix (this is the tricky part). Remember the definition of fecundity from Section 4.3.2: The fecundity, F_x, is the average number, per individual of age x alive at a given time step, of offspring censused *at the next time step*. So, fecundity of, say, two-year olds is the number of zero-years olds at time t produced by individuals who were two years old at time $t-1$. Look at the two-year olds at time $t-1$ in the figure. There are two "N_2"s for time $t-1$. Look at the one on the left, because we are now working on the pre-breeding census case. How many of their offspring become one-year olds at the next time step? Follow the arrows between "N_2" for time $t-1$ and "N_0" for time t. There are two arrows, labeled "m_2" and "S_0". This is because an average two-year-old produces m_2 newborns (m is fertility), of which $m_2 \cdot S_0$ are censused as zero-year olds at the next census. So, the fecundity of two-year olds is their fertility multiplied by the survival rate of zero-year olds:

$$F_2 = m_2 \cdot S_0$$

and similarly for other are classes:

$$F_1 = m_1 \cdot S_0$$
$$F_0 = m_0 \cdot S_0$$

Combining these formulas, we get a Leslie matrix for pre-breeding census, which is multiplied by the vector $N(t-1)$ to give $N(t)$:

$$\begin{bmatrix} N_0(t) \\ N_1(t) \\ N_2(t) \end{bmatrix} = \begin{bmatrix} m_0 S_0 & m_1 S_0 & m_2 S_0 \\ S_1 & 0 & 0 \\ 0 & S_2 & 0 \end{bmatrix} \begin{bmatrix} N_0(t-1) \\ N_1(t-1) \\ N_2(t-1) \end{bmatrix}$$

For the second case, assume that we census the population just *after* the annual reproduction (a "post-breeding" census). In Figure 4.7, concentrate on the columns of N_x to the right of the breeding points. A post-breeding census model assumes that there is no mortality between breeding and the subsequent census. Note that in this census, the youngest animals censused will be newborns, so the first age class (zero-year olds) in this case refers to a different set of individuals than in the previous case. The number of one-year olds will be the number of zero-year olds in the previous census times the survival rate of zero-year olds during the past 12 months: $N_1(t) = N_0(t-1) \cdot S_0$. So, the first survival we use in our Leslie matrix (second row, first column) should be S_0. This is different from the previous (pre-breeding census) case.

Next, we have to compute fecundities. The abundance of the first age class is N_0. The fecundity of, say, two-year-olds is the number of newborns at time t produced by individuals who were two years old at time $t-1$. Look at the two-year olds at time $t-1$ in the figure. There are two "N_2"s for time $t-1$. Look at the one on the right, because we are now working on the post-breeding census case. How many of newborns did these individuals produce at the next time step? Follow the arrows between "N_2" for time $t-1$ and "N_0" for time t. There are two arrows, labeled "S_2" and "m_2". This is because on average S_2 of these individuals survive until the next breeding season, and those who survive produce m_2 newborns (m is fertility). So, the fecundity of two-year olds is *their* survival rate from age 2 to age 3 (S_2) multiplied by their fertility (m_2):

$$F_2 = S_2 \cdot m_2$$

and in general,

$$F_x = S_x \cdot m_x$$

The Leslie matrix for post-breeding census, multiplied by $N(t-1)$ gives:

$$\begin{bmatrix} N_0(t) \\ N_1(t) \\ N_2(t) \\ N_3(t) \end{bmatrix} = \begin{bmatrix} S_0 m_0 & S_1 m_1 & S_2 m_2 & 0 \\ S_0 & 0 & 0 & 0 \\ 0 & S_1 & 0 & 0 \\ 0 & 0 & S_2 & 0 \end{bmatrix} \begin{bmatrix} N_0(t-1) \\ N_1(t-1) \\ N_2(t-1) \\ N_3(t-1) \end{bmatrix}$$

Note that this matrix has one more row and column than the matrix for pre-breeding census, since three-year olds are also observed. Note also that even though the last column is all zeros, we still keep it to tally the three-year olds.

For example, assume that we have estimated the following maternities (say from observations of fledglings per nest) and survival rates (say from a mark-recapture study), and that we are modeling only the female population.

$m_0 = 0.5 \quad S_0 = 0.3$
$m_1 = 2.5 \quad S_1 = 0.8$
$m_2 = 3.0 \quad S_2 = 0.5$

Now, if the initial abundances have been estimated in a pre-breeding census, then the Leslie matrix will become

$$\begin{bmatrix} 0.5 \times 0.3 & 2.5 \times 0.3 & 3.0 \times 0.3 \\ 0.8 & 0 & 0 \\ 0 & 0.5 & 0 \end{bmatrix} = \begin{bmatrix} 0.15 & 0.75 & 0.9 \\ 0.8 & 0 & 0 \\ 0 & 0.5 & 0 \end{bmatrix}$$

This matrix will predict, for each time step, the population size and structure just before breeding. You should keep this in mind when interpreting results, and also when deciding on a quasi-extinction threshold.

If, on the other hand, the initial abundances are estimated in a post-breeding census, the Leslie matrix becomes

$$\begin{bmatrix} 0.3 \times 0.5 & 0.8 \times 2.5 & 0.5 \times 3.0 & 0 \\ 0.3 & 0 & 0 & 0 \\ 0 & 0.8 & 0 & 0 \\ 0 & 0 & 0.5 & 0 \end{bmatrix} = \begin{bmatrix} 0.15 & 2.0 & 1.5 & 0 \\ 0.3 & 0 & 0 & 0 \\ 0 & 0.8 & 0 & 0 \\ 0 & 0 & 0.5 & 0 \end{bmatrix}$$

This matrix will predict, for each time step, the population size and structure right after breeding. This matrix is reducible as a result of a zero fecundity in the last element of the first row (see Caswell 1989). If you delete the last row and column the matrix becomes

$$\begin{bmatrix} 0.15 & 2.0 & 1.5 \\ 0.3 & 0 & 0 \\ 0 & 0.8 & 0 \end{bmatrix}$$

which has the same finite rate of increase as the two other matrices above. The difference is that, this will not allow the model to include the last age class (three-year olds) in the total population abundance. You may or may not want this; in this example, the three-year olds do not breed after they are counted in a post-breeding census, so it may be okay to exclude them from the total abundance.

4.7.3 Estimating Variation

Given a time series of estimates for a particular vital rate (say, zero-year-old survival rate), we can estimate the standard deviation of this vital rate using built-in functions in a calculator or spreadsheet software. However, this seemingly simple procedure may have many complications. In this section, we discuss two of these.

4.7.3.1 Variance Components

As we mentioned in Chapter 2 and again here in Chapter 4, elements representing the uncertainty in a population model must be estimated from data. In many instances, a single estimate of variation is available for each parameter. We know that this variation has several sources, but rarely are corrections made to identify the sources of the various components of the total variation. The total variance ($\text{var}_{\text{total}}$) in each parameter may be decomposed as:

$$\text{var}_{total} = \text{var}_{environment} + \text{var}_{sample} + \text{var}_{demography} + \text{var}_{space}$$

where $\text{var}_{environment}$ represents variation from year to year (or census to census) that is the result of the population's response to environmental variation, var_{sample} represents measurement error, $\text{var}_{demography}$ is variance due to demographic stochasticity, and var_{space} is the spatial variance among estimates that is due to measurements taken at different places. To these terms we could add covariances between each combination of sources.

If measurements from one year to the next are always taken in the same place, then var_{space} may be ignored. The act of sampling itself may affect the values that are likely to be recorded next time (the development of trap shyness in animals, for instance). For survival rates the term $\text{var}_{demography}$ is equal to the binomial variance, $p(1-p)/N$, where p is the survival rate and N is the number of individuals, so that if N is large (more than about 20), usually the term may be ignored. Sometimes, estimates of numbers come with an estimate of the associated measurement error. Assuming no covariance between sources, this would provide the means of reducing the total variance to its individual components.

4.7.3.2 Variance of Sums and Products

Often data will involve the sum of independent components, each of which has a variance associated with it. Such circumstances occur when creating composite classes, or when estimating abundance from spatially separate areas. Then, the variance (var_{1+2}) of the sum of two numbers is the sum of their respective variances plus 2 times their covariance (cov_{12}):

$$\text{var}_{1+2} = \text{var}_1 + \text{var}_2 + 2\,\text{cov}_{12}$$

In some circumstances, it may be necessary to estimate the variance of the product of two numbers, each of which has a variance associated with it. For example, when the data come in the form of maternities and survivorships, fecundity is given by

$$F = m \cdot S$$

The variance of the product of two values (1 and 2) is given by

$$\text{var}_{1 \times 2} = \text{var}_1 (\text{mean}_2)^2 + \text{var}_2 (\text{mean}_1)^2 + 2\,\text{mean}_1\,\text{mean}_2\,\text{cov}_{12}$$

4.8 Exercises

Before you begin this set of exercises, you need to know a few things about RAMAS EcoLab. For age- and stage-structured models, click on the icon for "Age and stage structure" from the RAMAS EcoLab main program (shell). This program works just like the previous program (Single population models), except for the larger number of parameters. See the Appendix at the end of the book for an overview of RAMAS EcoLab. For on-line help, press F1, double click on "Getting started" and then on "Using RAMAS EcoLab." You can also press F1 anytime to get help about the particular window (or, dialog box) you are in at that time. To erase all parameters and start a new model, select "New" under the Model menu (or, press Ctrl-N).

Note that the same program is used for both age-structured models and stage-structured models. In RAMAS EcoLab, the Leslie matrix of an age-structured model is entered in the **Stage matrix** (under the Model menu). The matrix is called a "Stage matrix," and the classes are called "stages" in RAMAS EcoLab, even if the classification is actually based on the age of organisms. This is because "stage" is a more general concept, and age-structured models can be considered as a special case of stage-structured models (we will discuss this in the next chapter). Therefore, many parts of the program refer to "stages," which you should assume to be "ages" for the exercises of this chapter. For example, in various windows, the classes are labeled by default as "Stage 1," "Stage 2," etc.; however, the actual meaning of stages depends on the particular model. For example, if the model is age-structured, "Stage 1" (i.e., the first age class) may refer to zero-year-old or one-year-old individuals, depending on the way your model is structured. Therefore, you should change the default labels to fit your model. This is done in the **Stages** dialog box, which is selected from the Model menu. Before you can enter the elements of the matrix, you must first decide on the number of age classes and enter each age class in the **Stages** dialog box (click the Help button for more information). There are two constraints that apply to age-structured models: all matrix elements must be nonnegative, and survival rates must be less than one. RAMAS EcoLab checks both of these, as long as the box for Ignore constraints in **General information**, also under the Model menu, is clear (not checked). This option should always be cleared for age-structured models.

When entering data for an age-structured model, make sure that the matrix in the **Stage matrix** (under the Model menu) has the structure of a Leslie matrix; fecundities should be entered in the top row, survival rates should be in the subdiagonal, and all other numbers should be zero. Make sure that survival rates are not on the diagonal by mistake.

144   Chapter 4 Age Structure

Exercise 4.1: Building the Helmeted Honeyeater Model

This exercise is designed to familiarize you with the program and to review the various concepts introduced in this chapter. We will begin by entering the basic parameters for the Helmeted Honeyeater matrix model we have discussed above.

Step 1. Start RAMAS EcoLab and select the program "Age and stage structure" by clicking on its icon.

Step 2. Start a new model. This will open the **General information** window. Type in appropriate title and comments (which should include your name if you are going to submit this work for assessment).

Enter the following parameters of the model. Remember that setting the number of replications to 0 is a convenient way of making the program run a deterministic simulation.

Replications:	0
Duration:	50
☐ Ignore constraints *(clear)*	

Note that the parameter related to demographic stochasticity is ignored. This is because when the number of replications is specified as 0, the program assumes a deterministic simulation. This parameter is ignored because it is relevant only for stochastic models. After editing the screen, click the "OK" button. (Note: Don't click "Cancel" or press [Esc] to close an input window, unless you want to undo the changes you have made in this window.)

Next, select **Stages** (again, under the Model menu). Click the "Add" button to increase the number of stages to 4. Click on the top cell under "Name." Change the default name, "Stage 1," by typing "Age 0." For the other rows, type "Age 1," "Age 2" and "Age 3+". The window should now look like

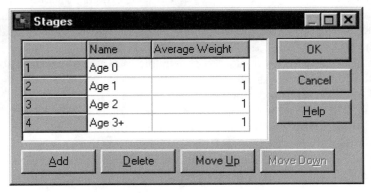

Click OK. Select **Stage matrix**. Click on the matrix element in the first column and second row, and type 0.703 (the survival rate of zero-year olds). Type the other elements of the matrix (survivals and fecundities) as estimated in Section 4.4, and displayed at the end of Section 4.4.1. When finished, the matrix should look like

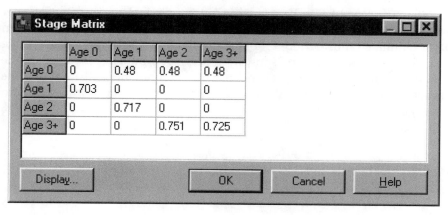

Click OK.

Step 3. Next, select **Initial abundances** and enter the abundances in 1994. Don't forget to combine the counts for ages 3 and above into a single age class (the numbers should be 29, 20, 14, 34). When finished, click OK, and save the model in a file.

Step 4. Select **Run** to run a simulation. The simulation will run for 50 time steps, and you will see "Simulation complete" at the bottom of the window when it's finished. For a deterministic simulation, this will be quite quick. Close the Simulation window, and select "Trajectory summary" from the Results menu. You will see an exponentially increasing population trajectory. Click on the second button from left ("show numbers") on top of the window to see the results as a table of numbers. The first column shows the time step, the others show five numbers that summarize the total abundance (of all age classes) for each time step: (1) minimum, (2) mean − standard deviation, (3) mean, (4) mean + standard deviation, and (5) maximum. All five numbers should be the same (because this is a deterministic simulation). Make a note of the last two, i.e., $N(49)$ and $N(50)$. Calculate the growth rate from year 49 to year 50:

$$R(49) = \frac{N(50)}{N(49)} = \underline{} = 1.\underline{}$$

This is an estimate of the finite rate of increase (λ) because (1) the model does not have any type of stochasticity or density dependence, and (2) the simulated time (50 years) is long enough that we can assume that the population has reached its stable age distribution. Now, let's check this last assumption.

Step 5. In RAMAS EcoLab, you can output the final age distribution after a simulation. This is the abundance of individuals in each age class at the end of the simulated time period. From the Results menu, select "Final age/stage abundances" and click the "show numbers" button. This table gives the abundance in each age class at the end of the simulation. Write down the numbers in each class, and then compute the proportion of individuals in each age class:

$$P_0 = \frac{N_0(50)}{N(50)} = \underline{\quad\quad} = 0.\underline{\quad\quad}$$

$$P_1 = \frac{N_1(50)}{N(50)} = \underline{\quad\quad} = 0.\underline{\quad\quad}$$

$$P_2 = \frac{N_2(50)}{N(50)} = \underline{\quad\quad} = 0.\underline{\quad\quad}$$

$$P_{3+} = \frac{N_{3+}(50)}{N(50)} = \underline{\quad\quad} = 0.\underline{\quad\quad}$$

Note that $N_x(50)$ is the number of x-year olds in year 50, and $N(50)$ is the number of all individuals in year 50. If the population has reached its stable age distribution, these numbers should be the same as the stable age distribution calculated based on the matrix. Let's check if this is the case. Close the result window and select **Stage matrix** from the Model menu. Click the "Display" button, and select "Finite rate of increase" by clicking on it. The program will display various statistics about this Leslie matrix in numerical form. Scroll down the window to see the value of the finite rate of increase (λ). Below that, various variables are tabled numerically. In addition to the stable age distribution and reproductive value distribution, here you can see the initial age distribution (i.e., the proportion of individuals in different age classes at the beginning of the simulation), and average residence times (we will discuss the average residence times in the next chapter). Notice that the initial age distribution is different from the stable age distribution.

Step 6. Compare the stable age distribution with the final age distribution you calculated, and the finite rate of increase with the growth rate from year 49 to year 50.

Step 7. Repeat steps 3 through 6 with a different age structure. Enter the initial abundances as 67, 10, 10, and 10 for the four age classes. Note that the total initial abundance is the same as in the previous case, but with more zero-year olds and fewer individuals in other age classes. Does the population reach stable age distribution in 50 years? Is the growth rate from year 49 to 50 close to the finite rate of increase? Now compare the final population size at year 50 for the two cases. Is the final abundance different when the initial distribution is skewed towards zero-year olds?

Step 8. Now we will add demographic stochasticity to the model. Load the file you saved in Step 3 and select **General information**. Notice that the number of replications is 0 (specifying a deterministic simulation). Change the number of replications to 50, and check the demographic stochasticity box. Run the model again. Note that each trajectory simulated by the program is different. If the simulation takes a very long time, you can speed it up by clicking on the first button on the toolbar of the simulation window. This button displays simulation text, instead of each trajectory (you can also stop a simulation by pressing (Esc) or clicking the "Cancel" button). After the simulation is over, select **Trajectory summary** from the Results menu. Although the only source of stochasticity is demographic, and the initial population size is 100, the expected future trajectory of the population shows considerable variation. Make a note of the variation in abundance. For example, for year 50, record the (1) minimum, (2) mean − standard deviation, (3) mean, (4) mean + standard deviation, and (5) maximum abundances.

Step 9. Now we will add environmental stochasticity to the model. Select **Standard deviation matrix** from the Model menu, and enter the numbers we calculated in Section 4.5.2. In this window, the standard deviation of each matrix element (survival and fecundity) is entered at the same position as in the **Stage matrix** screen. Thus, type in the standard deviation of zero-year-old survival (0.0364) in the second row, first column; type in the standard deviation of one-year-old fecundity (0.0075) in the first row, second column, etc. (Note that we arrange the standard deviations in the form of a matrix only for visual convenience; we don't do operations such as matrix multiplication with this matrix.)

When finished, click OK, and save the model in another file. Now, run a simulation. How does the variation compare with Step 8 (when we considered only demographic stochasticity)?

Step 10. The population abundance increased from 97 to over 1000 in 50 years. Considering the discussions in Chapter 3, what are some of the factors that might prevent such an increase? (Hint: See the beginning of Section 4.3.) One way to model such factors is to add density dependence to the model. Select **Density dependence** from the Model menu, and specify the type of density dependence as "Ceiling." This requires an additional parameter, the carrying capacity (K). Assume that K is 150. Click OK, and run another simulation. What does the population trajectory look like now? What is the long-term predicted abundance of the population? Does the finite rate of increase (based on the Leslie matrix) say anything about the population's future in the presence of stochasticity and density dependence?

Exercise 4.2: Human Demography

In earlier chapters, we discussed the exponential nature of human population growth, and the capacity of earth to support the human population. In this exercise, we will demonstrate some of the difficulties in dealing with human population growth.

Step 1. Load the file Human.ST, which is a model of the human population in a typical developing country. (The abundances and rates in this model are loosely based on the population of the Philippines in 1975.) The time step in this model is a decade, and the age structure is based on 10-year age classes. For example, the first class ("Stage 1") is ages 0–10, the second class ("Stage 2") is ages 10–20, etc. This is very important to remember during this exercise. Select **Stage matrix**, and click "Display" to investigate the vital rates. What is the finite rate of increase of this population? (Note that a finite rate of increase must be specified together with the time unit for which it was estimated.) Compare the initial age distribution with the stable age distribution. Are they the same?

Step 2. Simulate the growth of this population for 100 years. How many time steps does this take? What is the expected population size in year 2075? How much did the population increase (in absolute terms and as a percentage) in 100 years? Note that in this step we assumed that the 1975 fecundities and survivals remain unchanged for 100 years.

Step 3. Simulate the effects of a family planning program, for example one that makes birth control available free, and gives incentives for small families. Assume that the effect of this program is very strong and immediate; decrease the fecundity of each age class by the same percentage so that the finite rate of increase is equal to 1.000. What percentage decline in fecundity was necessary to make the long-term population growth zero (finite rate of increase = 1)? Save the model in another file.

Step 4. Simulate the population growth with the reduced fecundities until 2075. What will be the expected population size in year 2075? How much will the population increase (in absolute terms and as a percentage) in 100 years under the reduced fecundities? Why did the model predict that the population will continue to increase even though the finite rate of increase is equal to 1.0000?

Step 5. Repeat Step 3, but reduce fecundities in another way. Instead of reducing fecundities of all age classes by the same amount, begin with the youngest reproductive age class (ages 10–20). Decrease its fecundity until the finite rate of increase is equal to 1.0000. If the finite rate of increase is above 1.0 even when the fecundity of 10–20-year-old individuals is zero, start decreasing the fecundity of the next age class. Save the file under a new name. What percentage did you have to decrease the fecundity of each age class? Describe what this means in terms of the reproductive behavior of the people in this example? How is this sort of reduction in fecundities different from the one in Step 3?

Step 6. Repeat Step 4 with the new set of fecundities. Compare the results with those of Step 4? Which method results in a lower population size?

Step 7. How realistic is our assumption that the family planning program we simulated will decrease the finite rate of increase to 1.0 immediately? If this decrease takes a number of years (or decades), how would this affect the final population size? How might social factors (such as education of, and economic independence of women, and increased social security for older people) affect the rate with which fecundities decrease?

Step 8. Remember from Chapter 1 that per capita energy consumption in industrial countries is about 9.3 times that in developing countries. If in the next 100 years, the per capita energy consumption in the developing country in this example reached the level of consumption in the industrialized world, how much would the total annual energy consumption in that country increase by 2075:

(a) if there is no change in fecundities?
(b) if fecundities change as in Step 3?
(c) if fecundities change as in Step 5?

Exercise 4.3: Leslie Matrix for Brook Trout

Brook Trout (*Salvelinus fontinalis*) is a freshwater fish that is popular with anglers. Table 4.5 gives the number of brook trout in Hunt Creek (in Michigan), taken from a paper by McFadden et al. (1967). The data in this and the next table are provided on the distribution disk of RAMAS EcoLab in three spreadsheet formats (for Lotus 1-2-3, Quattro Pro, and Excel).

Step 1. Calculate the survival rates for each age group (S_0, S_1, S_2, and S_3) in each year. For example, S_0 for 1949 is 2013/4471, or 0.4502. Note that you cannot calculate the survival rates for 1962 (because there is no data for 1963). Also note that $S_4 = 0$, because no five-year olds were observed. If you know how to use spreadsheet software, see below before you begin this step.

Step 2. Calculate the average survival rate for each age class.

Table 4.5. Abundance of Brook Trout in Hunt Creek by age classes.

Year	Age classes					Total
	0	1	2	3	4	
1949	4,471	2,036	287	14	0	6,808
1950	3,941	2,013	304	13	0	6,271
1951	4,287	1,851	265	16	1	6,420
1952	5,033	1,763	261	16	0	7,073
1953	5,387	1,637	175	13	0	7,212
1954	6,325	2,035	234	13	0	8,607
1955	4,235	2,325	383	24	0	6,967
1956	4,949	1,612	392	51	1	7,005
1957	6,703	1,796	309	33	1	8,842
1958	5,097	2,653	355	26	2	8,133
1959	4,038	2,395	685	68	0	7,186
1960	5,057	2,217	473	47	1	7,795
1961	2,809	2,017	409	23	0	5,258
1962	5,052	1,589	448	52	2	7,143

From McFadden et al. (1967).

Step 3. Calculate the standard deviation of each survival rate. You can do this in three different ways. Any one of the three is acceptable (although they may give slightly different results).

(a) If you have (and know how to use) any one of the three spreadsheet software mentioned above, first load the file BTROUT.WK1 (for Lotus 1-2-3), BTROUT.WQ1 (for Quattro Pro), or BTROUT.XLS (for Excel). You can then calculate the survival rates (Step 1) by dividing the appropriate numbers (be careful with the years). After calculating the survival rates for each age and year, calculate their averages (Step 2) and their standard deviations (Step 3) using the built-in function of the software to calculate averages and standard deviations. Read the manual of the software you have for more information. If the software gives an option of either "population," or "sample" standard deviation, use the "sample standard deviation."

(b) If you have a calculator that performs standard deviation calculations, you can use it. Note also that the Calculator program that comes with Microsoft Windows also allows the calculation of standard deviations (when you select "View/Scientific"). Use the help facility of this program to learn how to use it.

(c) You can also use a short-cut that allows an approximate estimation of standard deviation from a range of observations. Range is the difference between the maximum and minimum of a set of numbers. To use this method, first calculate the range for each survival rate, by subtracting the minimum ever observed over the 13 years, from the maximum. Then divide this number by 3.336. This number is valid only for samples of 13 data points (as is the case here). For samples of different sizes, the constant used to divide the range is different (see Sokal and Rohlf 1981, page 58).

Step 4. Calculate the fecundities of each age class, and each year. For example, to calculate the fecundity of one-year olds (F_1) in 1949, you need to divide the number of zero-year olds alive in 1950 that were produced by one-year olds, with the number of one-year olds in 1949. From the above table, we know the number of one-year olds in 1949 (2,036), and the total number of zero-year olds alive in 1950 (3,941), but we need to know how many of these zero-year olds were the offspring of individuals that were one year old in 1949. We can obtain this information from the Table 4.6, which gives, for each year, the proportion of young produced by each age class (the sum of each row is 1.0). For example, in 1949, 64.7% of the young were produced by one-year olds. Thus the number of young (that were alive in 1950) that were produced by one-year olds in 1949 was 0.647 multiplied by 3,941, or 2,550. The fecundity of one-year olds in 1949 was

$$F_1(1949) = \frac{0.6471 \times 3941}{2036} = 1.2525$$

The fecundity of two-year olds in 1949 was

$$F_2(1949) = \frac{0.3193 \times 3941}{287} = 4.385$$

Note that both fecundities use the same total number of zero-year olds in 1950 (3,941). Calculate the rest of the fecundities. There should be a total of 52 fecundities (4 age classes, 13 years). Note that you cannot calculate the fecundity for 1962 (because there is no data for one-year olds in 1963), and that the fecundity of the first age class, $F_0 = 0$, i.e., this year's young cannot produce young that are counted in the next census. Also note that in 8 out of 13 years, the abundance was zero in age class 4. For these years you cannot calculate F_4 (it is *not* zero; it is unknown).

The data in this and the previous table are provided on the distribution disk of RAMAS EcoLab in three spreadsheet formats (for Lotus 1-2-3, Quattro Pro, and Excel). If you have (and know how to use) any one of these software, you can make these calculations much faster.

Table 4.6. Proportion of all young produced by different age classes of Brook Trout in Hunt Creek.

Year	Age classes			
	1	2	3	4
1949	0.6471	0.3193	0.0336	0
1950	0.6417	0.3333	0.0250	0
1951	0.6396	0.3063	0.0450	0.0090
1952	0.6275	0.3333	0.0392	0
1953	0.6750	0.2750	0.0500	0
1954	0.6827	0.2885	0.0288	0
1955	0.6096	0.3425	0.0479	0
1956	0.5000	0.3731	0.1269	0
1957	0.5726	0.3333	0.0940	0
1958	0.6358	0.2980	0.0596	0.0066
1959	0.4906	0.4151	0.0943	0
1960	0.5422	0.3675	0.0904	0
1961	0.5833	0.3681	0.0486	0
1962	0.4823	0.4043	0.1064	0.0071

From McFadden et al. (1967).

Step 5. Calculate the average fecundity for each age class. Calculate F_4 as the average of five numbers (three of which are zero), and other fecundities as averages of 13 numbers.

Step 6. Calculate the standard deviation of each fecundity. You can do this in three different ways (see above). Be careful when calculating the standard deviation of F_4. You should calculate the standard deviation of five numbers. For example, if you use the range approximation, divide the range of F_4 with 2.326 (because the sample size is 5, not 13).

Step 7. Combine the average survival rates and average fecundities into a Leslie matrix. Make another matrix with the corresponding standard deviations.

Exercise 4.4: Fishery Management

In this exercise, you are asked to manage a fishery. Your goal is to maximize the harvest, while minimizing the risk of decline. This fishery exercise is based on the brook trout model you developed in the previous exercise.

Step 1. Start RAMAS EcoLab, select the "age and stage structure" program. Select "New" from the File menu (this will open **General information**). Enter an appropriate title (and if you wish, comments), and enter the following parameters of the model:

Replications:	0
Duration:	20
☐ Ignore constraints *(clear)*	

Click OK. Select **Stages**, and click "Add." Rename the stages as "Age 0," "Age 1," etc. Click OK.

Step 2. Select **Stage matrix**, and enter the Leslie matrix you calculated in the previous exercise. Select **Standard deviation matrix**, and enter the numbers you calculated in the previous exercise. Select **Initial abundances**, and enter the abundance of each age class in 1962, from the table in the previous exercise. In each window, click OK after entering the parameters. Save the model in a file.

Step 3. Run a deterministic simulation of this model. Record the final population size (Trajectory summary; total abundance at year 20).

Step 4. Now, we will add harvesting. Two types of harvesting can be simulated with the program. Both are specified in **Management & Migration** under the Model menu. On the left side, there is list of management actions. Click the "Add" button under this list once to add a new management action. A new action will be added to the list on the left of this dialog box. The newly added action is assumed to be a "harvest/emigration."

First, we will simulate proportional harvest. On the right side of the window, under "Quantity" select "Proportion of individuals" by clicking on it. Then enter a number between 0 and 1 (say, 0.1) in the edit box next to the label "Proportion of individuals." This is the proportion of each age class harvested. Next, you need to select the age classes to which this harvest rate applies. We will assume that the zero-year-old fish are too small to be of commercial value, so we skip this first age class. The abundance of the last age class is too low to experiment with, so we will not harvest this class either. We will assume that the same proportion of other age classes are harvested. This may not be a valid assumption in most cases, but it does simplify the exercise. Click on the little arrow next to "In Stages," and select "Age 1." Click on the little arrow next to "Through," and select "Age 3." Click OK.

Step 5. Now change the harvest rate (i.e., the "Proportion of individuals") in such a way that the abundances in the last 10 years are as close to each other as possible (i.e., the population is stationary). You will probably have to run several simulations before you can find the correct number that keeps the population sizes stationary. What is the harvest rate you found? Save this model in a different file (such as TROUT-PH.ST, for proportional harvest). Record the number of fish in each age class at the end of the simulation (Final stage age/stage abundances).

Step 6. We will now repeat Steps 4 and 5 with constant harvest for the same age classes. Constant harvest refers to a fixed number of individuals harvested at each time step. But we don't want this number to be the same for all age classes (because there are many more younger fish than older fish). To guess these constant numbers, multiply the number of fish in each age class at the end of the simulation (from the previous step) with $h/(1-h)$, where h is the proportional harvest rate you found, then round to the nearest integer. For example, suppose the abundance of the second stage ("Age 1") at the end of the simulation in Step 5 was 2,000, and the harvest rate was 0.06. In this case use a constant harvest of $2,000 \times 0.06/(1-0.06)$, or 128 fish for this age class. Select "Management & Migration" and click on "Number of individuals" (under "Quantity"). Type "128" as the number. Make sure that this number applies only to "Age 1." Thus, click on the little arrow next to "Through," and select "Age 1." Thus this management action refers to harvesting 128 individuals in stages "Age 1" through "Age 1."

Calculate the number to be harvested for the other two age classes in the same way. To enter the number for "Age 2," click the "Add" button. This adds a new management action to the list, also named "Harvest/Emigration." Click on the newly added "Harvest/Emigration." The numbers on the right side of the window now refer to this new action. Enter the number you calculated as the "Number of individuals," and change "In stages ... through ..." to refer to "Age 2." Repeat for "Age 3." Thus, you should have three managament actions, all of the "harvest/emigration" type, and each refering to a single age class. Click OK.

Run a deterministic simulation, and check the final abundance. If the population is increasing or declining, adjust the constant harvest numbers (proportionally) until the abundances in the last 10 years are as close to each other as possible (i.e., the population is stationary). You probably won't have to make any adjustments to the initial guesses.

What are the harvest amounts you found? Save this model in a different file (such as TROUT-CH.ST, for constant harvest).

Step 7. Now run stochastic simulations (by changing the number of replications to 1000, and making sure demographic stochasticity is used) with each of the three models (no harvest, proportional harvest, and constant harvest) you have developed and saved. Check the risk of falling below 1000 individuals for each simulation. (It might be difficult to read the precise value of the probability from the screen plot. See Exercise 2.4 for an example of getting the exact probability value.)

Compare the three results. Explain the differences in the light of the discussion on "Harvesting and density dependence" in Chapter 3.

4.9 Further reading

Jenkins, S. H. 1988. Use and abuse of demographic models of population growth. *Bulletin of the Ecological Society of America* 69:201–202.

Keyfitz, N. and W. Flieger. 1990. *World population growth and aging: demographic trends in the late twentieth century.* University of Chicago Press, Chicago.

Leslie, P. H. 1945. On the use of matrices in certain population mathematics. *Biometrika* 33: 183–212.

McFadden, J. T., G. R. Alexander and D. S. Shetter. 1967. Numerical changes and population regulation in brook trout *Salvelinus fontinalis*. *Journal of the Fisheries Research Board of Canada* 24: 1425–1459.

U.S. Bureau of the Census home page. http://www.census.gov/
Includes demographic data on the population of the U.S. and other countries.

Chapter 5
Stage Structure

5.1 Introduction

The basic assumption of age-structured models is that the demographic characteristics of individuals (such as fertilities and survival chances) are related to their age, and among individuals of the same age, there is little variation with respect to these demographic characteristics. This assumption is not appropriate for all species; age is not always a good indicator of demography. In some plants, survival and reproduction depend on the size of the individual. Larger individuals produce more seeds and are more likely to survive. Such a species could be modeled with an age-structured model, only if individuals in the same age class were more or less the same size. Usually this is not true; plant growth is often plastic, meaning that the rate with which individual plants grow in size depends on environmental conditions. Those seeds that happened to land on a favorable spot will grow faster and reproduce at an earlier age than those that were less lucky. Forest trees, for example, can spend years suppressed in the understory before an opening in the canopy allows them to grow and begin producing seeds. Openings in the canopy occur when canopy trees die due to chance events such as wind, disease, and fire. As a result, the age of a tree that has been waiting in the understory may have nothing to do with its chance to begin growing to the canopy.

In such a case, it may be better to use a stage-structured model, in which the individuals are grouped into stages defined by their physiological, morphological, or other characteristics that have an important effect on their probability of survival and reproduction. For the above example, the stages may include seeds, seedlings, saplings, understory trees, and canopy trees.

Animal species may also be better modeled using a stage-structured approach. For example, survival rates and fecundities may depend on the physiological stages—such as egg, larva, pupa, and adult for insects, or juvenile and adult for birds. Again, such species could also be modeled with a Leslie matrix, but only if individuals took the same length of time to reach these stages (in other words, if each individual spent the same length of time in each stage). Otherwise, age structure will not capture the differences among individuals in terms of their survival and reproduction.

There might also be practical reasons for using stage structure instead of age structure. For example, it may be impossible to determine the age of individuals, hence impossible to estimate age-specific vital rates. In such cases, a stage-structured model may be more appropriate.

5.2 Assumptions of stage-structured models

The basic assumption of stage-structured models is that the demographic characteristics of individuals are related to their developmental stage. The assumption is that there is little variation among individuals in the same stage with respect to their demographic characteristics such as chance of surviving, chance of reproducing, and the number of offspring they produce.

This assumption is quite important. It means that what an organism will do depends only on the stage it is in *now*, and not on what stage it was in the previous time steps, or how long it remained in each stage. For example, a stage-structured model of forest trees (mentioned above) based on size would assume that the chances of survival and growth of an individual sapling depend on its size, but not on how long it has waited in the understory, or whether it was a seed or seedling in the previous time step.

Other than this basic assumption, stage-structured models may also assume that (1) the population is closed, i.e., there is no immigration or emigration; (2) the vital rates are constant, i.e., there is no demographic or environmental stochasticity; (3) the vital rates are not dependent on abundance, i.e., there is no density dependence. However, it is quite easy to dispense with these assumptions and add migration, stochasticity, and density dependence to a stage-structured model.

5.3 Stage structure based on size

The most important difference between an age-structured and a stage-structured model involves the type of transitions; in other words, the number and type of transitions possible for an individual in a given age or stage class. In an age-structured model, there are only two types of transitions: an individual may get older (i.e., move to the next class), and/or it may produce offspring (i.e., contribute to the first age class). If neither of these happens, the model assumes that the individual died. These two types of transitions are represented by the two types of nonzero elements of the Leslie matrix: those at the subdiagonal (i.e., elements one below the diagonal going from the upper left of the matrix to the lower-right), representing survival, and those in the first row, representing fecundities. Below is the Leslie matrix we discussed in the previous chapter, with the assumption that $F_0 = 0$, i.e., the youngest individuals do not reproduce (which is often the case). We can depict this age-structured model with a diagram (Figure 5.1) in which boxes represent age classes and arrows represent transitions (survivals and fecundities) from one age class to another.

$$L = \begin{bmatrix} 0 & F_1 & F_2 & F_3 \\ S_0 & 0 & 0 & 0 \\ 0 & S_1 & 0 & 0 \\ 0 & 0 & S_2 & 0 \end{bmatrix}$$

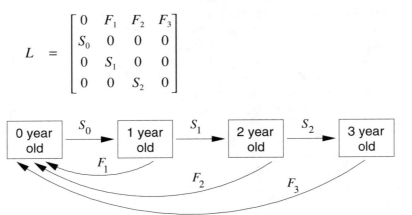

Figure 5.1. Diagram for an age-structured model.

We discussed one modification of the Leslie matrix in the previous chapter involving a third type of transition. When we combined three-year-old and older individuals into a composite class, their survival rate was represented by the matrix element at the lower-right corner of the matrix below. We will use the symbol S_{3+} for the survival of three-plus-year-old individuals. When these individuals survive for another year, they are still counted in the same class. This is represented by the loop around the box for "3+ year old" in Figure 5.2. The rest of the figure is the same as Figure 5.1.

160   Chapter 5 Stage Structure

$$L = \begin{bmatrix} 0 & F_1 & F_2 & F_{3+} \\ S_0 & 0 & 0 & 0 \\ 0 & S_1 & 0 & 0 \\ 0 & 0 & S_2 & S_{3+} \end{bmatrix}$$

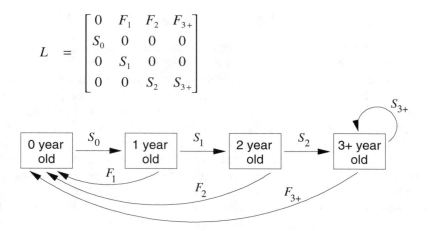

Figure 5.2.   Diagram for an age-structured model with a composite age class for individuals three-years-old and older.

Now consider a model in which individuals in any class (not just the last one) can either move to the next class or stay where they are. Obviously this cannot happen in an age-structured model. If there is an age class for two-year-olds, then a one-year-old individual will either die, or survive to become two years old. However, if the classes are based not on age, but on the size of the individuals, then a "medium-sized" individual can grow to be "large-sized," or stay as "medium-sized." Such a model may be represented by a diagram with loops for each stage (Figure 5.3), or by the following stage matrix.

$$L = \begin{bmatrix} S_{TT} & F_{ST} & F_{MT} & F_{LT} \\ S_{TS} & S_{SS} & 0 & 0 \\ 0 & S_{SM} & S_{MM} & 0 \\ 0 & 0 & S_{ML} & S_{LL} \end{bmatrix}$$

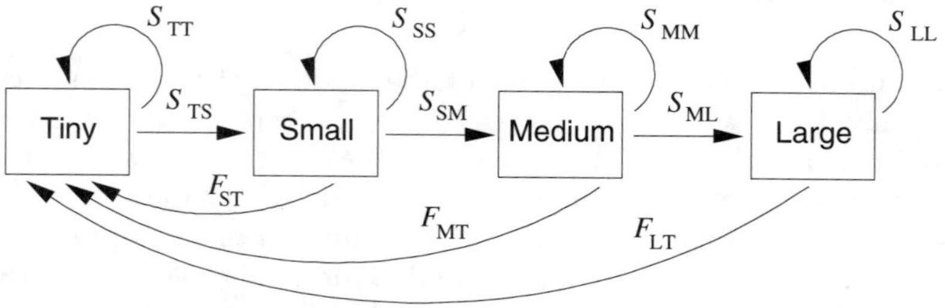

Figure 5.3.   Diagram for a stage-structured model.

In this matrix, the subscripts T, S, M, and L refer to "tiny," "small," "medium," and "large." All survival rates now have a subscript with two letters, indicating the beginning and ending stages referred to by the survival rate. For example, S_{ML} is the proportion of "medium" individuals who became "large" in the next time step, and S_{MM} is the proportion of "medium" individuals who remained as "medium" in the next time step. The overall survival rate of "medium" individuals is

$$S_M = S_{MM} + S_{ML}$$

and their fecundity is F_{MT}, because fecundity is a transition from one stage (in this case "medium") to the stage in which individuals start their lives. In this model we assumed that all offspring start their lives as "tiny" individuals. This is not necessarily the case for all species. There may be small and big offspring (seedlings, for example), and you might want to model them in separate classes. We also assumed that a "small" individual can become a "large" individual in two time steps, because it must first become a "medium" individual. If it were possible for a "small" individual to become "large" in a single time step, then (1) there would be a nonzero element S_{SL} in the second column, last row of the matrix, and (2) there would be an arrow going directly from "small" to "large" in Figure 5.3. Note that, in these diagrams, each element of the matrix is represented by an arrow; the number of arrows in a diagram is equal to the number of nonzero elements of the matrix.

5.4 A stage model for an Alder

A stage-structured model based on size was developed by Huenneke and Marks (1987) for the Speckled Alder (*Alnus incana*). This is a common shrub of eastern North America. It forms dense thickets in which alder seedlings have a very low survival rate; thus most of the reproduction is vegetative, in the form of sprout production.

Huenneke and Marks (1987) censused and measured alders from 1979 to 1982. They classified alders in their study populations with respect to the diameter of their stems at breast height (dbh; diameter at 1.4 m above the ground), a common measure of size for trees and shrubs. They grouped stems into the following five size classes:

Stage 1: 0 cm dbh (i.e., stems shorter than 1.4 m)
Stage 2: 0.1 to 0.9 cm dbh
Stage 3: 1.0 to 1.9 cm dbh
Stage 4: 2.0 to 2.9 cm dbh
Stage 5: 3.0 to 3.9 cm dbh

Because the census was repeated every year from 1979 to 1982, Huenneke and Marks (1987, Table 2) estimated three stage matrices, each representing a transition from one year to the next (1979 to 1980, 1980 to 1981, and 1981 to 1982). We combined these three matrices and obtained the following mean matrix.

	1	2	3	4	5
1	0.637	0.033	0.100	0.163	0.230
2	0.107	0.590	0.0	0.0	0.0
3	0.0	0.353	0.763	0.0	0.0
4	0.0	0.0	0.237	0.667	0.0
5	0.0	0.0	0.0	0.277	0.737

In this matrix, the first row (except for the first element, 0.637) refers to vegetative production of new sprouts, the diagonal elements (0.637, 0.59, ..., 0.737) refer to the proportion of stems that survive and remain in the same size class, and the subdiagonal elements (0.107, ..., 0.277) refer to the proportion of stems that survive and increase in size to the next class. For example, the fates of individuals in stage 2 are given by the numbers in the second column of the matrix: on average, 59% of stems in stage 2 remain in the same stage after a year, and 35.3% of them increase in size. The rest (1−0.59 − 0.353 = 5.7%) die. The first number in this column indicates that on average each stem in this class produces 0.033 sprouts. This means that many stems do not produce any sprouts, and the total number of sprouts produced by stage 2 stems, divided by the number of stems in this stage is, on average, 0.033.

For stage 5, there is a single number for survival. All stage 5 stems that survive remain as stage 5 stems, because this is the stage for the largest stems. The largest stems happened to have the largest fecundity (an average of 0.23 sprouts per stem).

As we mentioned above, Huenneke and Marks estimated three matrices for three years. We calculated the above mean matrix as follows. For each element, we calculated the arithmetic average of the corresponding matrix elements from the three matrices (Huenneke and Marks discuss other methods of combining data from three years). For example, the survival rate of stems in stage 5 was estimated as

0.83 from 1979 to 1980
0.71 from 1980 to 1981
0.67 from 1981 to 1982

The average of these numbers is 0.737 (the number in the above matrix), and their standard deviation is 0.068. We can calculate a standard deviation for each element of the matrix, because we have three estimates (from three years) for each element. We can arrange the standard deviations we have calculated in the form of a matrix, so that the standard deviations will correspond to the means in the above matrix. (Note that, as we discussed in the previous chapter, we do this only for visual convenience; one cannot do operations such as matrix multiplication with this matrix). The result is the following matrix of standard deviations.

	1	2	3	4	5
1	0.118	0.009	0.029	0.046	0.067
2	0.066	0.139	0.0	0.0	0.0
3	0.0	0.188	0.071	0.0	0.0
4	0.0	0.0	0.071	0.066	0.0
5	0.0	0.0	0.0	0.078	0.068

Given a mean stage matrix and the corresponding standard deviations, the calculations needed to make projections for a population are very similar to the calculations we discussed in the previous chapter. As in the age-structured models, we also need to know the initial number of individuals in each stage, arranged in the form of a vector (a column of numbers, one number for each stage). To make a projection, we multiply a stage matrix with this vector, as we did for the age-structured model of Helmeted Honeyeater in the previous chapter. In a stochastic model, the matrix we use for this projection is not the mean matrix, but a different matrix at every time step. We select the elements of this matrix from random distributions with the means and standard deviations given in the two matrices above. We then make the matrix multiplication

$$N(t+1) = M(t) \cdot N(t)$$

where $M(t)$ is the stage matrix for year t, and $N(t)$ is the vector of stage abundances in year t.

5.5 Building stage-structured models

When biologists build stage-structured models for the species they study, they often start by deciding how to divide the population into stages. In the previous sections, we discussed models based on size of individuals. In a stage-structured model, individuals in a population may be grouped into classes based on characteristics other than size. This may be the physiological, morphological, or developmental state of the individuals, or a combination of one of these with size or weight. How the population is divided into stages depends on several factors. The most important factor is what the demography of the species depends on. If survival rates or fecundities have nothing to do with the size of an individual, then there is no point in building a size-structured model. Other factors may involve more practical considerations such as the ease of identifying different life stages or measuring other characteristics, and the amount of data available.

For an insect species, a matrix model might group individuals into developmental stages such as egg, larva, pupa, and adult. For a bird species, the stages might be juveniles (or fledglings), non-breeding adults, and breeding adults.

The difference between age-based and stage-based matrix models is that in a stage matrix, any element can be greater than zero (though stage matrices with *all* elements greater than zero are quite rare). The element at the *i*th row and *j*th column of a stage matrix represents the rate of transition from stage *j* to stage *i*. Rate of transition in general means the proportion of individuals that were in stage *j* at time *t* that "become" stage *i* individuals at time *t*+1, or "contribute to" the number of stage *i* individuals at time *t*+1 through reproduction. Thus, each element can describe reproduction, survival, or both.

Consider the following stage-structured model of Jack-in-the-pulpit (*Arisaema triphyllum*), a perennial herb of deciduous forests (Bierzychudek 1982, Table 2: the matrix shows the transitions in the Fall Creek population from 1977 to 1978).

	1	2	3	4	5	6	7
1	0	0	0.07	1.82	4.69	6.51	7.00
2	0.20	1.17	0.56	0.49	0.47	0.47	0.47
3	0	0.10	0.60	0.06	0.06	0.13	0
4	0	0	0.04	0.68	0	0.07	0
5	0	0	0.03	0.09	0.12	0	0.14
6	0	0	0.01	0.09	0.29	0.27	0.14
7	0	0	0.01	0.04	0.41	0.53	0.71

The stages are seeds (stage 1) plus six size classes based on leaf area (stages 2 through 7). The first row of this matrix represents the fecundity (seed production) of plants in different stages, and the first column gives the survival rate of seeds. There is only one number in this column, thus surviving seeds (20% of all seeds) become the smallest plants. In the second column, we have two numbers, just as we did in the size-structured model of the previous section. The second number (0.10) is the rate of transition from stage 2 to stage 3. Note, however, that the rate of transition from stage 2 to stage 2 is greater than 1.0 (unlike in the previous section's model). Thus, it cannot refer only to the proportion of stage 2 plants that remain as stage 2 plants (this would be less than or equal to 1.0). The reason it is greater than 1.0 is that this number also includes vegetative reproduction by stage 2 plants. When there is vegetative reproduction, the "offspring" (i.e., the recruits to the population) are plants rather than seeds. In this case these new individuals enter the population in stage 2.

How about the third column? This column shows what happens to stage 3 individuals. The first number in this column (0.07) is again the seed production. The second number is, again, the sum of two numbers: vegetative reproduction by stage 3 individuals, and the proportion of stage 3 individuals that become stage 2 individuals in the next time step. Because the stages are defined in terms of plant size, this means that some part of this number (0.56) represents plants that become smaller in size. This is not unusual in models based on size. This column shows that Jack-in-the-pulpit may have very plastic growth. While some plants may decrease in size, others (about 1% of stage 3 plants) may grow a lot, to reach the largest size class in just one year.

5.5.1 Residence Times, Stable Distribution, and Reproductive Value

Another property of stage-structured models that is different from those with age structure concerns the average time individuals spend in each stage. In an age-structured model this is the same for all age classes. If age is defined in years, each individual spends exactly one year in each age class (the only exception is the composite age class, if there is one). In a stage-structured model, individuals may spend different amounts of time in different stages. The *average* time individuals spend in a stage is given by the reciprocal of one minus the diagonal element:

$$\text{Residence time in stage } i = \frac{1}{(1 - S_{ii})}$$

This assumes that the diagonal element does not include reproductive transitions. If it does, the reproductive rate must be subtracted before the above formula is used. If the diagonal element is zero, then residence time is 1 time step (as is the case for age-structured models). If the diagonal element is 0.5, then this means that half the individuals remain in the same stage, and the average residence time becomes 2 time steps.

In the previous chapter on age-structured models, we discussed three variables that are based on the Leslie matrix: the finite rate of increase (λ), the stable age distribution, and the reproductive value distribution. These variables have the same meaning (and are calculated in the same way) for stage-structured models. The stable stage distribution gives the proportion of individuals in each stage that all initial distributions converge to, if given enough time (assuming no stochasticity, migration, or density dependence). The finite rate of increase (λ) is the rate with which the population grows once it reaches the stable distribution. Reproductive value gives the number of offspring an individual in a given stage will produce, relative to those produced by an individual in the first stage.

5.5.2 Constraints

In an age-structured model, elements of the first row of the matrix represent fecundities, and elements in the other rows represent survival rates. Most stage-structured models have a similar characteristic. Usually, there is only one (often, the first) stage that represents new recruits to the population (such as stage 1, for "sprouts," in the Speckled Alder model discussed above). This means that only the first-row elements represent fecundities; elements in the other rows are survival rates (either surviving and moving to another stage or surviving in the same stage). If this is the case, the sum of all elements in a given column, excluding the first row, is the total proportion of survivors from that stage, and cannot exceed 1.0. (Note that the Jack-in-the-pulpit example provides an exception, which we will discuss below.)

For example, in the Speckled Alder model, 66.7% of stage 4 plants in a given year remain in the same size class in the following year, and 27.7% grow to the next size class (stage 5). Thus, a total of 94.4% of stage 4 plants survive and the rest (5.6%) die. Obviously, the total of these two transitions (stage 4-to-stage 4 and stage 4-to-stage 5) in this model must be less than or equal to 100%. This is especially important in stochastic models, where sampling the transition rates from random distributions may result in column sums of above 1.0, even if the mean values were restricted to be between 0 and 1.

In RAMAS EcoLab, you can specify whether you want to impose constraints on matrix elements to ensure that column sums (excluding the first row) are always between 0 and 1. As a default, the first row of the matrix is assumed to represent reproduction, and all other rows are assumed to represent survival rate. The program checks both the average matrix (specified in **Stage matrix** under the Model menu) during editing, and each sampled stage matrix during a simulation, to make sure that (1) all elements are non-negative, (2) for each column, all elements except the one in the first row add up to less than or equal to 1.0. If the first check fails, negative elements are set to zero. If the second check fails, the program makes automatic corrections, by proportionally decreasing each nonzero element (except the first element) of the column until the sum equals 1.

In some cases, this constraint is invalid because there is recruitment to stages other than the first one. This means that elements in rows other than the first one also represent reproduction (perhaps vegetative reproduction), and therefore should not be restricted to add up to 1.0 (for example, the Jack-in-the-pulpit model). If this is the case, you must check the box labeled **Ignore constraints** in **General information**. The program then will ignore the second check discussed above; and any element, as well as any column sum, may be above 1.0 (although all elements are still constrained to be nonnegative).

5.5.3 Adding Density Dependence

When we added various types of density dependence to our models in Chapter 3, the models did not have age or stage structure. The population growth was determined by the growth rate (R) and its variation. We modeled density dependence by making the average value of R a function of abundance.

In the previous chapter, the only density dependence we added to an age-structured model was of the ceiling type. This type of density dependence is simple to add to any model, because it works in a similar way whether there is age/stage structure or not. In either case, if population abundance increases above the ceiling (K), then it is decreased back to K. The only difference in the case of models with age or stage structure is to decide which age or stage class abundance to decrease. One option is to decrease abundances of all ages or stages proportionally (this is what RAMAS EcoLab does). For example if $K = 1,000$, and $N = 1,040$, then abundance in each age class or stage is decreased by 40/1040, or by 3.8%. As long as the abundance is below the ceiling, the population grows (or declines, or fluctuates) according to the stage matrix and the standard deviations.

When density dependence is of the scramble or contest type (such as the models in Section 3.8.1), then the model becomes more complicated. In this case, we want the population's average growth rate to be a function of abundance. But in an age- or stage-structured model, the growth rate is not a specific parameter; it is a result of various parameters (survivals, fecundities) that make up the Leslie matrix or the stage matrix. Abundance is also not a single variable; it is made up of the abundances of the different age classes or stages. Because of multiple parameters (fecundities and survivals) that might be affected by the abundance, and because of different measures of abundance, there are many different ways of modeling density dependence in an age- or stage-structured model. What we use in RAMAS EcoLab is one of the simplest ways: the total abundance (of all ages/stages) affects all elements of the stage matrix (fecundities and survivals) proportionally.

This density dependence is implemented in RAMAS EcoLab in such a way that the result is the same as in the simpler density-dependent models we discussed in Chapter 3. When the total abundance (of all ages/stages) is equal to K (when the population is at its carrying capacity), then the growth rate of the population (determined by the stage matrix) is 1.0; when the population is above its carrying capacity ($N > K$), the growth rate becomes less than 1.0, and when the population is below its carrying capacity ($N < K$), the growth rate increases above 1.0. When the population abundance is so low that the effects of density dependence are negligible, then the average growth rate is equal to R_{max}, the maximum rate of increase. R_{max} is a required parameter (in addition to K) if the density dependence is of the scramble or contest type.

5.6 Sensitivity analysis

An important question that comes up in studies involving stage-structured models is the contribution of each matrix element to the dynamics of the population. In other words, how sensitively does the population's future depend on each element of the stage matrix. There may be several reasons for asking this question. Two of the common reasons involve planning of future field research and evaluating management options.

5.6.1 Planning Field Research

Because of lack of sufficient data and measurement errors, parameters of a model are often known as ranges instead of single estimates. For example, we may know that the average juvenile survival is between 0.30 and 0.60, and the average adult survival is between 0.85 and 0.9, but may not know exactly what the averages are. In such cases, collecting more data makes these ranges narrower, and consequently the results become more certain. But if there are many parameters (transitions among many stages) that are known with such uncertainty, which should we try to estimate better first? Given that there is a cost associated with additional field work, it makes sense to know whether our research money is better spent collecting data for, say, juvenile survival or adult survival.

There are three considerations in making such a decision. The first one is the contribution of each parameter (each matrix element) to population growth. There are various methods for making this calculation. Some of these methods, such as "sensitivities" and "elasticities," are based on the effect of each vital rate on the eigenvalue of (finite rate of increase given by) the stage matrix. These measures are reported in RAMAS EcoLab (click "Display" in **Stage matrix**, select "Sensitivities and elasticities," and scroll down the window; press F1 for additional information). However, these measures ignore variability, density dependence and the initial distribution of individuals to stages (the program gives a warning about the relevant factors ignored by these measures). In addition, they focus on the deterministic growth rate, rather than the more relevant results such as the risk of extinction.

Another method of calculating sensitivities involves calculating the effect of each matrix element on the risk of extinction or chance of recovery of the population. This is similar to the sensitivity analysis described in Exercise 2.4 (Chapter 2), in which we changed each parameter by plus and minus 10%, and checked the difference in the probability of an increase with the low and the high value of each parameter. The advantage of this method is that it incorporates all the factors in the model (including density dependence and variability), and it focuses on probabilistic results (extinction risk or recovery chance).

The second consideration in deciding which parameters are more important to estimate more precisely is the uncertainty in each parameter. For example, if we are very uncertain about juvenile survival (e.g., the estimated range is 0.3 to 0.6) and reasonably certain about adult survival (e.g., the estimated range is 0.85 to 0.90), then it would make sense to spend more time and money for additional data on juvenile survival. With the risk-based method described above, we can take this consideration into account by changing each parameter to the lower and upper values of its estimated range (i.e., 0.3 and 0.6 for juvenile survival; 0.85 and 0.90 for adult survival), instead of changing them plus and minus a fixed percentage. This way, a parameter with a wider range will contribute more to uncertainty about the risk of extinction (other things being equal).

With the deterministic methods (such as elasticities), it is not always possible to take this consideration into account, because those methods are based on linear approximations, which means they assume that growth rate changes linearly with changes in vital rates. This is often a good approximation for small changes, but may not be valid for large ones (e.g., when a survival rate is known as a wide range).

Another disadvantage of the deterministic methods is that they are often applied only to matrix elements. However, as we saw in the last chapter, some matrix elements may have to be estimated as products of two vital rates. For example, fecundity may be estimated as the product of maternity (e.g., number of fledglings per adult) and survival of the juveniles until the next census. If we want to decide whether the field work should focus on maternity or juvenile survival (which may require different types of study design), then the sensitivity of the population growth rate to their product (fecundity) is not very useful. In an exercise below, we will explore the sensitivity of the risk of decline of a spotted owl population to uncertainties in two different vital rates (each of which contribute to two stage matrix elements).

The third consideration is the relative cost of obtaining enough data for different parameters. For example, if it requires much more money to reduce the estimated range in, say, survival by a certain amount, than to reduce the range in fecundity by the same amount, it makes sense to focus on fecundity instead of survival. This consideration can be taken into account by first calculating the expected decrease in uncertainty in each parameter with a fixed amount of research money, and then using these ranges in the analysis. For example, we might guess that if we spend a certain amount of money obtaining more data on juvenile survival, we might reduce its range from [0.3–0.6] to [0.4–0.5], and with the same amount of money, we might reduce the range in adult survival from [0.85–0.90] to [0.86–0.89]. Obviously, such a guess would be approximate at best. Also, once a certain amount of data is collected, and new parameters are calculated, the relative contributions of each parameter will change. At that point, we will need to recalculate our strategy.

These considerations can also be extended to parameters other than those in the stage matrix (average vital rates). Often, the variabilities of vital rates are known even more poorly than their averages. We may be uncertain about the type of density dependence, or the number of stages to use in the model. In addition, factors such as density dependence may have strong effects on extinction risks. The risk-based sensitivity analysis that we first explored in Exercise 2.4 is suitable for incorporating parameters of a model other than the stage matrix elements. In each of these instances the strategy is the same: change model values or model structure to their alternatives and measure the importance of the change by the effect it has on the risks of decline.

5.6.2 Evaluating Management Options

Another application of sensitivity analysis involves decisions about which vital rates to focus on in management and conservation efforts. For example, protecting nests of Loggerhead Sea Turtles may increase the average fecundity, whereas installing escape hatches in shrimp trawl nets reduces the mortality of larger turtles (see Exercise 5.3 below). The decision about which conservation measure to invest in, is partly a question of whether it is better to increase fecundity or survival.

The evaluation of management options requires considerations similar to those for planning field research. The first is the contribution of each vital rate to the expected growth rate, and the chances of decline or recovery of the population. Thus, a formal sensitivity analysis of a model can provide some insight into how best to manage a population. If competition, for example, affects juveniles but juvenile survival contributes little to the growth rate or decline risk, then controlling species that compete with juveniles is unlikely to be of much help.

The second consideration is how much each vital rate (or other model parameter) can be changed with management. For some species, it may be possible to increase fecundity by, say, 10%, but adult survival (being already high) can perhaps be increased only by 5%. For some species, it may not be possible or practical to increase certain vital rates at all. Further complicating this issue is the fact that each management or conservation action may affect more than one vital rate. For example, protecting nest locations of a bird species may improve fecundity, and to a lesser extent survival rates, whereas restoring dispersal habitat may improve dispersal rates, juvenile survival, and to a lesser extent adult survival. In these cases, a parameter-by-parameter analysis of sensitivity does not make sense, because the parameters cannot be changed independently (or in isolation from others). It is much better to do a whole-model sensitivity analysis and compare management options instead of single parameters. This can be done by developing models for each management or conservation alternative. Each model incorporates changes to all the parameters affected by that particular alternative. The results of these models than can be compared to each other, as well as to a "no-action" scenario.

The third consideration is the relative cost of each management action. Even if, say, increasing adult survival by 5% results in a lower extinction risk than increasing fecundity by 10%, if the former is so expensive that, with the available resources, it can be carried out in fewer populations or for fewer threatened species than the latter, then perhaps the latter is the better option. In an exercise in Chapter 7, we will further explore the effect of cost on evaluating management options for an endangered bird species.

5.7 Additional topic

5.7.1 Estimation of Stage Matrix

Estimation of a stage matrix from data is similar to that of a Leslie matrix, with a few important differences.

The first step in determining the stage matrix is to decide on what the stages are. This mostly depends on the life history of species studied. If the stages are defined on the basis of the size of organisms, then the number of stages, and the size limits for each stage must also be decided. This may be a complicated problem. On the one hand, it is necessary to define a sufficiently large number of stages so that the demographic characteristics of individuals within a given stage are similar. On the other hand, it is necessary to have a sufficiently large number of individuals in each stage so that the transition probabilities can be calculated with reasonable accuracy (see Vandermeer 1975 and Moloney 1986).

Once the stages are defined, the estimation of the stage matrix elements depends on the type of data available. If individuals can be followed through at least two time steps, and their stage at each time step recorded, these data can be used in estimation by the following method, discussed by Caswell (1989). At each time step, individuals are identified by their stage. Since each individual's stage in the previous time step is also known, it can be assigned to a particular cell in the table below. The numbers in the cells represent the number of individuals making such a transition. Suppose such tallying for a particular time step yielded the following hypothetical table.

		At time $t-1$, individuals that were in stage:			
		1	2	3	4
At time t, individuals that are now in stage:	1	3			
	2	4	15		
	3		8	12	
	4		1	3	4
Deaths		3	6	5	12
Total		10	30	20	16

According to this example, out of the 10 individuals that were in stage 1 last year, 3 of them are still in stage 1 this year, 4 of them are now in stage 2 and the remaining 3 died. After all the individuals are thus tallied, non-reproductive transitions are calculated by dividing each stage-by-stage cell by the column total (which includes deaths). For example, 4 out of 10 in the above example corresponds to a transition rate of 0.4 from stage 1 to stage 2 per year. This calculation yields a four-by-four matrix. For this case, we get

$$\begin{bmatrix} 0.30 & 0 & 0 & 0 \\ 0.40 & 0.50 & 0 & 0 \\ 0 & 0.27 & 0.60 & 0 \\ 0 & 0.03 & 0.15 & 0.25 \end{bmatrix}$$

Note that if there are no individuals in a particular stage at time $t-1$, transition rates from that stage (i.e., the elements in the corresponding column of the stage matrix) cannot be estimated. In such a case, data from several time steps must be used to ensure that all columns have at least one positive element.

The calculation so far is based on following individuals, which allows estimation of transition from one stage to another. But a stage matrix also includes reproduction. To estimate fecundities with the same method, we need data on the number of new individuals added to each stage. Often new individuals are added only to the first stage (such as "seeds" or "fledglings"), but in some models they may be added to more than one stage (such as "small juveniles" and "large juveniles"). In addition to the number of such new individuals, we need data on the stage their parents belonged to. If we have such data, we can construct a table comparable to the one above. In this table, the number of recruits to stage i that are born of parents in stage j are recorded in cell i,j (i.e., row i, column j). For this example, suppose there were 40 recruits (to stage 1) at time t, all produced by parents who were in stage 4 at time $t-1$. Thus, all reproduction in this case is recorded in row 1 column 4 of the matrix. The number of offspring (40) is divided by the number of individuals in stage 4 at time $t-1$ (in this case, 16). The matrix of reproductive transition rates is therefore

$$\begin{bmatrix} 0 & 0 & 0 & 2.50 \\ 0 & 0 & 0 & 0 \\ 0 & 0 & 0 & 0 \\ 0 & 0 & 0 & 0 \end{bmatrix}$$

The nonreproductive transitions and reproductive transitions are then added together element-wise to obtain the following stage matrix for time $t-1$

$$\begin{bmatrix} 0.30 & 0 & 0 & 2.50 \\ 0.40 & 0.50 & 0 & 0 \\ 0 & 0.27 & 0.60 & 0 \\ 0 & 0.03 & 0.15 & 0.25 \end{bmatrix}$$

There will be a different matrix estimated for each time step, and means and standard deviations can be estimated for each matrix element. If the sample sizes differ greatly for different time steps, it might be necessary to calculate weighted averages for the transition probabilities (see Additional topics in Chapter 4).

Another way of estimating parameters for a stage matrix involves censusing the population at several time steps. At each census, all individuals are counted, and classified according to stage. This method does not require following each individual, but requires more years of data. It involves a multiple regression analysis for each stage (see Additional topics in Chapter 4 for multiple regression for age 0). For more information on methods of estimating the stage matrix, see Caswell (1989).

5.8 Exercises

The use of RAMAS EcoLab to build a stage-structured model is very similar to its use for age-structured models. For both, we use the same program ("Age and Stage Structure"). The main differences are: (1) in **Stage matrix** and **Standard deviation matrix**, any number can be greater than zero, and (2) in **General information**, the option Ignore constraints may be either checked or unchecked (clear), depending on the model (see Section 5.5.2 above).

Exercise 5.1: Reverse Transitions

The diagram for a stage structured model was provided in Figure 5.3. Consider the situation in which you observe an individual decrease in size between one census and the next. The change in size is sufficient that the individuals should be classified into a smaller size class. Such events may result from herbivory, or wind damage (in plants), or from loss of condition in animals. This would require new arrows going (say) from large to medium, or from medium to small. Assume that transitions of this kind were observed in Alder. Assume that 5% of all individuals in stage 3 were classified in the next census as stage 2 instead of stage 3 (in other words, instead of staying in the same stage, they moved to a smaller stage). The proportion that became larger did not change. Make the same assumptions for individuals in stages 4 and 5. Rewrite the matrix for Alder in Section 5.4, incorporating these new rates.

Exercise 5.2: Modeling a Perennial Plant

The following matrix is a stage-structured model of the Teasel (*Dipsacus sylvestris*), which is a perennial plant that is found mostly in disturbed habitats (Werner and Caswell 1977; Caswell 1989). The time step of the model is one year, and the stages in the model are
    (1) first-year dormant seeds (S1)
    (2) second-year dormant seeds (S2)
    (3) small rosettes (R1)
    (4) medium rosettes (R2)
    (5) large rosettes (R3)
    (6) flowering plants (FP)

	S1	S2	R1	R2	R3	FP
S1	0	0	0	0	0	322.380
S2	0.966	0	0	0	0	0.000
R1	0.013	0.010	0.125	0	0	3.448
R2	0.007	0	0.125	0.238	0	30.170
R3	0.008	0	0	0.245	0.167	0.862
FP	0	0	0	0.023	0.750	0.000

Notice that there are transitions from flowering plants to first-year dormant seeds, and all three classes of vegetative rosettes. In this case, these transitions do not represent vegetative reproduction, but the fact that seeds produced by flowering plants in one year may germinate to produce rosettes in the following year. This requires a transition from flowering plants to rosettes. (A transition from flowering plants to seeds, and seeds to rosettes would take two years instead of one year.) Because of this characteristic of the model, the constraint of keeping column sums less than or equal to 1.0 does not apply. Thus, in **General information**, Ignore constraints must be checked. This matrix was estimated from an experimental study in which each field in the study area was seeded with 3,900 teasel seeds in the winter. Thus the initial population consisted of 3,900 individuals (dormant seeds) in the first stage, and none in other stages. The experiment lasted for 5 years. We will use this as the simulation duration. This is appropriate because this species is often found in ephemeral habitats.

Step 1. Draw a diagram of this model.

Step 2. Enter the model into RAMAS EcoLab. In **Stage matrix**, click "Display" and select each type of graph, to answer the following questions.
(a) What is the most abundant stage at the stable stage distribution?
(b) Which stage has the highest reproductive value?
(c) On average, in which stage do individuals spend the most time?
(d) Is the initial distribution similar to the stable distribution?
(e) What is the annual rate of increase?
(f) Calculate the number of individuals you would expect in the population in one year and in two years, based only on this growth rate, and the initial abundance of 3,900.

Step 3. Run a deterministic simulation for 5 years. What is the population size in year 2? How does it compare with your prediction (in Step 2) based only on the growth rate? Why is there a difference?

Step 4. We do not know the variation in stage matrix elements, so in this exercise we will assume that there is only demographic stochasticity. Remember that demographic stochasticity is especially important in small populations. Do you think that this model will give very different results when you add demographic stochasticity?

Step 5. Run a simulation (using demographic stochasticity) with 1,000 replications. How do population projections compare to the deterministic projection?

Step 6. What is the probability that this population will exceed 20,000 individuals (including dormant seeds) anytime within the next 5 years?

Exercise 5.3: Sea Turtle Conservation

Loggerhead Sea Turtle (*Caretta caretta*) is a threatened marine reptile. It is a long-lived iteroparous species. Determining the age of Loggerhead Sea Turtles is very difficult due to their fast juvenile growth and their brittle shell that cannot hold marking tags. The following stage matrix is from a study by Crowder et al. (1994). In this matrix, the time step is one year, and the stages are defined as follows:

(1) hatchlings
(2) small juveniles
(3) large juveniles
(4) subadults
(5) adults

	1	2	3	4	5
1	0	0	0	4.665	61.896
2	0.675	0.703	0	0	0
3	0	0.047	0.657	0	0
4	0	0	0.019	0.682	0
5	0	0	0	0.061	0.8091

In this matrix, the two numbers in the first row represent the fecundity of subadults and adults. The diagonal elements (for example, 0.703 for small juveniles) specify the proportion of the individuals in a stage this year that will be in the same stage in the following year. The subdiagonal elements specify the proportion of individuals in that stage that grow to the next stage in the following year (for example, 4.7% of the small juveniles become large juveniles each year). The sum of diagonal and subdiagonal elements give the total rate of survival for individuals in that stage (e.g., 75% of small juveniles survive per year). Thus, in **General information**, Ignore constraints must be unchecked (which is the default). We will make the following assumptions about this model:

(1) The initial abundance is 100,000 turtles, distributed among stages as 30,000 hatchlings, 50,000 small juveniles, 18,000 large juveniles, and 2,000 subadults (no adults).

(2) The standard deviation of each vital rate (each element of the matrix) is 10% of its mean value.

(3) The density dependence is Ceiling type, and the carrying capacity is 500,000 turtles.

Step 1. Enter the model into RAMAS EcoLab, and save it in a file. Run a simulation (with both demographic and environmental stochasticity) for 30 years, and report the following:

(a) Probability of increasing to more than 200,000 turtles sometime in the next 30 years.

(b) Probability of falling below 20,000 turtles sometime in the next 30 years.

It might be difficult to read the precise value of the probability from the screen plot. See Exercise 2.4 for an example of getting the exact probability value.

Step 2. One of the threats the Loggerhead Sea Turtle faces is accidental capture and drowning in shrimp trawls. One way to prevent these accidents is to install escape hatches in shrimp trawl nets. These are called turtle exclusion devices (TED); they can drastically reduce the mortality of larger turtles (i.e., large juveniles, subadults, and adults). The following matrix shows what might happen to the stage matrix if TEDs were widely installed in existing trawl nets.

	1	2	3	4	5
1	0	0	0	5.448	69.39
2	0.675	0.703	0	0	0
3	0	0.047	**0.767**	0	0
4	0	0	**0.022**	0.765	0
5	0	0	0	**0.068**	0.876

The numbers in bold show the vital rates that are assumed to increase as a result of TEDs. Both the proportion remaining in the stage and the proportion growing to the next stage are higher for the three stages affected. In addition, the fecundities are slightly higher. This is because the fecundities give the number of hatchlings this year, per subadult/adult turtle in the previous year. Thus fecundities incorporate both fertility, and the survival rate of subadults and adults. If subadults and adults survive better, then fecundity is also higher.

Enter the model with TEDs into RAMAS EcoLab, and save it in a different file. Keep all other parameters (including the standard deviations) the same. Repeat the simulation as in Step 1, and report the following:

(a) Probability of increasing over 200,000 turtles sometime in the next 30 years

(b) Probability of falling below 20,000 turtles sometime in the next 30 years

How would TEDs change the prospects for this species?

Step 3. Another important source of mortality for most marine turtles occurs in the very beginning of their lives, between the time the eggs are laid in a nest in the beach, and the time they hatch and are able to reach a safe distance into the sea. Most turtle conservation efforts in the past have concentrated on enhancing egg survival by protecting nests on beaches or removing eggs to protected hatcheries (Crowder et al. 1994). We will assume that the effect of such an effort is an increase in the fecundity values.

Load the first turtle model you created (without the effect of TEDs). Your goal is to find out how much the fecundities must increase to give the same probability of increasing over 200,000 turtles as the model with TEDs (in Step 2). Increase the two fecundities by the same proportion (any proportion), and run a simulation (do not change the standard deviations). Check the probability of increasing over 200,000 turtles sometime in the next 30 years. If it is less than what you found in Step 2, increase them some more (again, in proportion). If it is more, decrease the fecundities. You don't need to get exactly the same answer. If the two probabilities (with TEDs and with beach protection) are within 0.1 of each other, you can stop.

How much must the beach protection increase fecundity in order to offer the same protection to turtles as offered by TEDs? (In other words, What is the ratio of the final fecundity to the unchanged fecundity?) Which method seems more effective?

Exercise 5.4: Sensitivity Analysis

Northern Spotted Owl (*Strix occidentalis caurina*) is a threatened species inhabiting the old-growth forests of the northwestern United States. Demographic studies on various populations of this subspecies have been summarized by Burnham et al. (1996). The following stage-structured model is based on data from one of these studies (in Willow Creek study area in northwest California; "CAL" in Burnham et al. 1996).

This model assumes a birth-pulse population and a post-reproductive census (see the section on "Estimating a Leslie matrix from a life table" in the previous chapter). In this model, there are three stages: juveniles are newly fledged owls, subadults are one year old, and adults are all older owls. The following are the model parameters.

S_j : juvenile survival rate; the proportion of fledglings that survive to become one-year old subadults

S_s : subadult survival rate, proportion of one-year old subadult owls that become two-year olds

S_a : adult survival rate, proportion of older owls that survive one year

m_j, m_s, and m_a : maternities (number of fledglings produced per owl) of juveniles, subadults, and adults, respectively

Note that because of the assumption of post-reproductive census, m_j refers to the number of fledglings produced by an owl that has fledged in the previous census, almost a year ago. Thus m_j is the maternity of owls that are almost 12 months old, and S_j is the survival of fledglings to become subadults. The product $S_j \cdot m_j$ is the number of fledglings produced by each juvenile that was counted in the last year's census (see Figure 4.7 in the previous chapter). Thus, the stage matrix is

$$\begin{vmatrix} S_j \cdot m_j & S_s \cdot m_s & S_a \cdot m_a \\ S_j & 0 & 0 \\ 0 & S_s & S_a \end{vmatrix}$$

The following table gives the values of these parameters for the study area "CAL," together with their standard error (Burnham et al. 1996). Standard error (S.E.) is a measure of the measurement or sampling error associated with the estimation of each parameter.

	m_j	m_s	m_a	S_j	S_s	S_a
Mean	0.094	0.205	0.333	0.33	0.868	0.868
S.E.	0.067	0.077	0.029	0.043	0.012	0.012

In this exercise, we will use the *standard errors* as measures of parameter uncertainty (i.e., measurement error). In addition to this uncertainty due to measurement errors, the parameters also have natural variability due to environmental fluctuations. We will model environmental stochasticity with the following *standard deviations*:

$$\begin{vmatrix} 0.0294 & 0.0437 & 0.0711 \\ 0.0190 & 0 & 0 \\ 0 & 0.0499 & 0.0499 \end{vmatrix}$$

The method of calculating the standard deviations for this model is based on Akçakaya and Raphael (1998).

In this exercise, we will perform a sensitivity analysis, with the aim of deciding on which parameters to concentrate in future demographic studies. We will do this by analyzing how much the uncertainty in each survival rate contributes to the uncertainty of the results. In other words, which parameters are important in terms of the reducing the uncertainty in the model?

Step 1. In **General information** of RAMAS EcoLab ("Age and Stage Structure"), specify 1,000 replications and 50 time steps (years). Also, make sure that (1) "Use demographic stochasticity" is checked and (2) "Ignore constraints" is clear (not checked). In **Stages** name three stages as "Juveniles," "Subadults," and "Adults." In **Standard deviation matrix**, enter the standard deviations given above. In **Initial abundances**, enter 46, 41, and 313 for juveniles, subadults and adults, respectively.

Step 2. Calculate the stage matrix given above using the average estimates of the parameters, enter in **Stage matrix** and save in a file (e.g., named "NSOaverage").

Step 3. Create four additional models, with plus or minus 1 standard error of juvenile and adult survival. The four models will differ only in their stage matrix. For each model, calculate the stage matrix as described below (note that each survival contributes to two elements of the stage matrix). Do not change the mean maternity values.

(1) Juvenile survival = average *minus* one standard error. For all other parameters, use the average values.

(2) Juvenile survival = average *plus* one standard error. For all other parameters, use the average values.

(3) Adult survival = average *minus* one standard error. For all other parameters, use the average values.

(4) Adult survival = average *plus* one standard error. For all other parameters, use the average values.

Save each model in a separate file, with descriptive names (such as "HighAdultSurv," "LowJuvSurv," etc.).

Step 4. Run each model. Click the "text" button in the upper-left corner of the Simulation window to complete the simulations faster. Record the risk of falling to or below 50 individuals (i.e., risk of decline at threshold = 50) in the table below. Calculate the difference in risk with the low and high value of each parameter.

Probability of declining to 50

Parameter:	with low value	with high value	difference
Juvenile survival			
Adult survival			

Step 5. Which parameter needs to be estimated more precisely? Open the file you have saved in Step 2, with average values of all parameters. In **Stage matrix**, click first "Display," then "Sensitivities and elasticities," scroll down to the elasticity matrix? "Elasticities" and "Sensitivities" are measures of the contribution that each matrix element makes toward the dominant eigenvalue of the stage matrix (see Section 5.6). According to this result, which survival rate is more important? Is there a difference in the results of risk-based sensitivity analysis you performed, and the deterministic elasticities? If so, what might be the reason(s) for the difference?

5.9 Further reading

Caswell, H. 1989. *Matrix Population Models: Construction, Analysis, and Interpretation.* Sinauer Associates, Sunderland, Massachusetts.

Crowder, L. B., D. T. Crouse, S. S. Heppell, T. H. Martin. 1994. Predicting the impact of turtle excluder devices on loggerhead sea turtle populations. *Ecological Applications* 4:437–445.

Lefkovitch, L. P. 1965. The study of population growth in organisms grouped by stages. *Biometrics* 21:1–18.

Usher, M. B. 1966. A matrix approach to the management of renewable resources, with special reference to selection forests. *Journal of Applied Ecology* 3:355–367.

Chapter 6
Metapopulations and Spatial Structure

6.1 Introduction

In the previous chapters, we developed models with varying degrees of complexity. In developing each of these models, we focused on the dynamics of a single population. This is often sufficient as many of our questions concern populations within confined areas, such as the extinction risk of the Helmeted Honeyeater in a single nature reserve (Chapter 4), management of the Brook Trout fishery in a river (Chapter 4), and the growth of the Muskox population on an island (Chapter 1). In other cases, the population may live in a large area, but the relative uniformity of its habitat suggests the use of a single population model, as in the case of the Loggerhead Sea Turtle

184 Chapter 6 Metapopulations and Spatial Structure

(Chapter 5). But often, species exist in a number of populations that are either isolated from one another or have limited exchange of individuals. Such a collection of interacting populations of the same species is called a *metapopulation*. Each distinct population in a metapopulation may be referred to as a subpopulation, a local population, or simply as a population.

In developing models for species that live in more than one population, we need to address the interaction between these populations. For example, populations of Mountain Sheep (*Ovis canadensis*) in southern California inhabit mountain "islands" in a desert (Figure 6.1). These populations live in 15 of these mountain ranges, which are separated by 6 to 20 km of unsuitable desert habitat (Bleich et al. 1990). Mountain Sheep cannot live for long in the desert, but they can migrate through it. Bleich et al. (1990) documented movement of Mountain Sheep between 11 pairs of these mountain ranges, and concluded that the movement of sheep among mountain patches was important for their conservation for both genetic and ecological reasons.

Populations of many species like the Mountain Sheep occupy patches of high-quality habitat and use the intervening habitat only for movement from one patch to another. Metapopulations both occur naturally as a result of spatial heterogeneity, and are created as a result of human actions. We will discuss these two factors next.

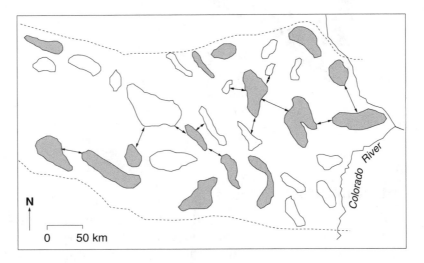

Figure 6.1. Populations of Mountain Sheep in Southern California. Shaded areas indicate mountain ranges with resident populations, arrows indicate documented intermountain movements; the dotted lines show fenced highways (after Bleich et al. 1990).

6.1.1 Spatial Heterogeneity

Spatial heterogeneity refers to the nonuniform distribution or occurrence of environmental variables and events in different parts of the landscape. Many species naturally exist as metapopulations because the environmental factors necessary for their survival occur in patches. For example, Giant Kelp (*Macrocystis pyrifera*) in the coastal waters off southern California grow in "forest" patches determined mostly by the properties of the substrate on the ocean floor, exposure to wave action, and water depth (Burgman and Gerard 1989). There are numerous other examples of patchy distribution of habitats; you may think of ponds in a forest, islands in an archipelago, or mountain ranges in a desert.

One of the assumptions we made with the first set of models (Chapter 1) is that all individuals, no matter where they occur, experience the same changes in the environment and the same chances of surviving and reproducing. This assumption is not valid for most metapopulations. In addition to the spatial variation in environmental factors (such as soil conditions, elevation, vegetation, water depth, etc.), many of the extreme events we discussed in Chapter 2 (such as fires, droughts, and floods) affect different populations of a metapopulation to varying degrees. The changes caused by such events are usually not uniform throughout the landscape, and how an individual fares will depend on where it happens to be. For example, fires often burn in mosaics that depend on fuel loads, moisture conditions, landscape characteristics, and prevailing winds. Different parts of the habitat burn at different intensities and with different frequencies, and some parts escape fire altogether.

Another example of the spatial heterogeneity of environmental factors is the disturbance pattern that characterizes the dynamics of Furbish's Lousewort (*Pedicularis furbishiae*), an endangered plant endemic to northern Maine (USA) and adjacent New Brunswick (Canada). It was assumed to be extinct for 30 years until its rediscovery in 1976 (Menges 1990). It is now known to exist in 28 populations along a 140-mile stretch of the St. John River.

The dynamics of the Furbish's Lousewort metapopulation are characterized by frequent extinctions of its populations caused by local disturbances such as ice scour and bank slumping, which are distributed patchily (Menges 1990). These disturbances also seem to be essential for the species' survival since they prevent tree and shrub establishment (events which would depress Lousewort populations). As a result, individual populations are short-lived, with fairly rapid increases followed by catastrophic losses. This natural disturbance pattern makes the viability of the species dependent on dispersal and establishment of new populations (Menges and Gawler 1986; Menges 1990).

6.1.2 Habitat Loss and Fragmentation

Loss of habitat is probably the most important cause of species extinction in recent times. Habitat loss often results not only in an overall decrease in the amount of habitat, but also in discontinuities in the distribution of the remaining habitat. Discontinuities can be created by opening land to agriculture, and by construction of buildings, dams, roads, power lines, and utility corridors. The result is the fragmentation of the original habitat that now exists into disjunct patches. Any population that inhabited the original habitat will now be reduced to a smaller total size and would be divided into multiple populations. Further fragmentation results in a decrease in the average size of habitat patches and makes them more isolated.

Other effects of fragmentation are manifested through increased edge effects. When habitat patches decrease in size through fragmentation, the populations inhabiting them become more vulnerable to adverse environmental conditions that are prevalent at the edges of the habitat patch, but not in its interior. For a forest patch embedded in an agricultural or a disturbed landscape, these environmental changes might include increased light and temperature or decreased humidity. They might also include biotic factors.

An example for biotic factors is the Brown-headed Cowbird (*Molothrus ater*) that parasitizes nests of forest-dwelling bird species. Cowbirds are more abundant at forest edges. They lay their eggs in the nests of other birds, who then raise cowbirds instead of their own young. In a study of forest fragmentation in the American Midwest, Robinson et al. (1995) found that parasitism by cowbirds was higher in more fragmented landscapes. This is because the proportion of forest that is away from the edges is lower in a forest that is made up of smaller patches. For example, if the interior of a forest that was not subject to edge effects such as cowbird parasitism began at 250 m away from the forest edge, then a hypothetical, circular patch of forest with a total area of 5 km^2 would have 64% interior forest habitat (Figure 6.2).

A patch with the same size edge but with an area of 1 km^2 would have 31% interior forest habitat, and a patch with 0.5 km^2 total area would have only 14% interior forest habitat. A 5 km^2 patch may seem to have 10 times the habitat as a 0.5 km^2 patch, but considering edge effects, it might actually have 46 times more habitat [$(5 \times 0.64)/(0.5 \times 0.14) = 45.7$]. Thus a landscape that has many small patches of habitat may have much less interior habitat than a similar-sized landscape with larger patches, because of edge effects. If the patches have shapes different from a perfect circle, or if the edge effects can penetrate a greater distance into the forest, this ratio would be even higher.

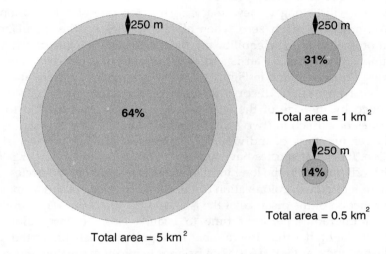

Figure 6.2. Edge effects: circular patches with an edge of 250 m, and areas of 5, 1, and 0.5 km^2. The percentages represent the ratio of the area of interior forest habitat (darker shaded regions) to the total patch area.

6.1.3 Island Biogeography

Island biogeography is concerned with the patterns of species richness on oceanic islands. One of these patterns is related to the size of the island. Larger islands often have more species than smaller islands. The relationship between the size of an island and the number of species it has is described by a species–area curve, which is often plotted on a graph with both axes in logarithms. The equilibrium theory of island biogeography (MacArthur and Wilson 1967) attempts to explain this pattern based on two processes: extinction and colonization. According to this theory, rate of extinction increases as more species are added to an island (solid line in Figure 6.3). Note that extinction rate here refers to the number of species that become extinct per unit time, and not to the extinction risk of any particular species. Assuming the extinction risks are constant, the number of species becoming extinct should increase with increased number of species on the island.

The other process, colonization of the island by new species, is expected to decrease as the number of species on the island increases (dashed line in Figure 6.3). Assuming that the rate of immigration is constant, when the number of species already present on the island is higher, fewer of the immigrants will belong to new species. Thus the rate of colonization will decrease, and will reach zero when all the species on the mainland are present on the island. The equilibrium theory of island biogeography views the number of species on an island as an equilibrium between these two processes. Thus the island in Figure 6.3 has S^* number of species at equilibrium. According to the theory, the rate of extinction of species is also determined by the size of the island; larger islands are expected to have lower rates of extinction (two solid lines in Figure 6.4). Larger islands may have larger numbers of individuals per species, causing a lower risk of extinction. If several species have lower risk of extinction on larger islands, those islands would, in the long term, have a larger number of species.

The other process, colonization of the island by new species, is expected to be a function of the distance of the island to the mainland. The number of species immigrating per unit time to islands close to the mainland is expected to be higher than the number immigrating to islands farther away (two dashed lines in Figure 6.4). The balance between extinction and colonization determines the equilibrium number of species on distant and large islands (S^*_{DL}), distant and small islands (S^*_{DS}), close and large islands (S^*_{CL}), and close and small islands (S^*_{CS}).

The pattern of increased number of species on larger islands has also been observed for habitat islands, i.e., patches of one type of habitat (say, forest) surrounded by another (e.g., agricultural areas). There are explanations for these patterns other than the equilibrium theory discussed above. For example, larger islands often have a greater variety of habitats, which contributes to the larger number of species.

Whether the equilibrium theory of island biogeography correctly determines the number of species on islands has been the subject of debate (see Burgman et al. 1988 for a review). Another concern from a conservation point of view is that the theory cannot be used to determine which species are likely to become extinct. Extinction risk of a species is determined to a large extent by factors other than those in the theory. In earlier chapters we examined some of these factors, such as stochasticity and density dependence. In the rest of this chapter, we will discuss factors that are important determinants of extinction risk at the metapopulation level.

Introduction 189

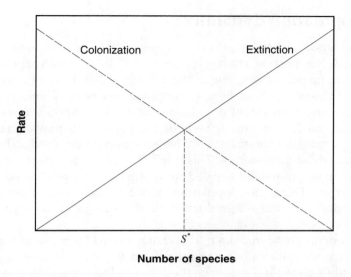

Figure 6.3. The rate of extinction and the rate of colonization as a function of the number of species present on an island, according to the equilibrium theory of island biogeography (MacArthur and Wilson 1967).

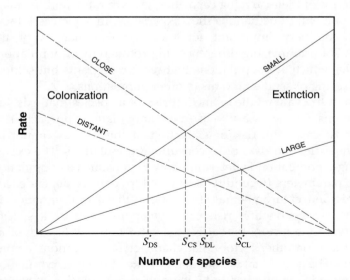

Figure 6.4. The rate of extinction and the rate of colonization for large, small, close and distant islands, according to the equilibrium theory of island biogeography (MacArthur and Wilson 1967).

6.2 Metapopulation dynamics

In previous chapters, we developed several types of models that aimed at evaluating the risk of extinction of a single population. Because many species live in more than one population, one of three things might happen after a single population becomes extinct. First, the population may be colonized by individuals dispersing from extant populations of the same species (extant means "surviving," or "not extinct"). Second, the population may remain extinct; in other words, the habitat patch where the population used to live remains unoccupied by that species. This might happen if the population lived in a remote patch of habitat isolated from other habitat patches occupied by the species. Third, the population may be recolonized through human intervention—by the reintroduction of the species to its former habitat.

The type of population dynamics that is characterized by frequent local extinctions and recolonizations is a natural pattern for many species (e.g., Andrewartha and Birch 1954). Thus, even though each local population may exist for only a short period of time, the metapopulation may persist for a long period, with a constantly changing pattern of occupancy of local populations in the metapopulation as patches "blink" off and on.

This dynamic complexity is further enriched by differences among the populations in terms of their carrying capacities, growth rates, and the magnitudes of environmental fluctuations they experience. In some cases these differences may be very important for the overall dynamics of the metapopulation. For example, big differences in productivity of populations may lead to sinks, which are populations that receive migrants but seldom produce any offspring or send emigrants to other populations.

When we want to evaluate the extinction risk of a species that exists in multiple populations, it is necessary to use a metapopulation approach. This is because in most cases the risk of extinction of the species cannot be deduced from the extinction risks of its constituent populations. The extinction risk of a single population is determined by factors such as population size, life history parameters (fecundity, survivorship, density dependence), and demographic and environmental stochasticity that cause variation in these parameters. The extinction risk of a metapopulation or a species depends not only on the factors that affect the extinction risk of each of its populations, but also on other factors that characterize interactions among these populations. The additional factors that operate at the metapopulation or species level include the number and geographic configuration of habitat patches that are inhabited by local populations, the similarity of the environmental conditions that the populations experience, and dispersal among populations that may lead to recolonization of locally extinct patches. We will discuss these factors next.

6.2.1 Geographic Configuration

When a species lives in several patches, much depends on exactly where those patches are, i.e., on their spatial arrangement. This determines the dispersal rates, as well as the similarity of the environmental conditions in neighboring patches (we will discuss these factors in the next two sections).

Metapopulation models assume that some parts of the landscape are habitat patches (that are, or at least can potentially be, occupied by populations), and the remainder is unsuitable habitat. In some cases, the species in question has a specific habitat requirement that has sharp boundaries, making patch identification quite straightforward. Most examples of patchy habitats we discussed above (ponds in a forest, islands in an archipelago, woods in an agricultural landscape, or mountaintops in a desert) fit this category.

In other cases, habitat quality varies on a continuous scale and designation of areas as habitat and nonhabitat may be somewhat arbitrary. Or, the boundaries may not be clearcut for human observers. What seems to be a homogeneous landscape may be perceived as a patchy and fragmented habitat by the species living there. If the suitability of habitat for a species depends on more than one factor, and some of these factors are not easily observable, the habitat patchiness we observe may differ from the patchiness from a species' point of view. An example is the Helmeted Honeyeater (*Lichenostomus melanops cassidix*) that we discussed in Chapter 4. The habitat requirements of this species include the presence or absence of surface water, the density of *Eucalyptus* stems, and the amount of decorticating bark on these stems (Pearce et al. 1994). In such cases, the information about habitat requirements may be combined by computer maps of each required habitat characteristic, using geographic information systems (Akçakaya 1994). This allows us, in effect, to see the habitat patches as perceived by the species. Akçakaya et al. (1995) used this approach to model a metapopulation of Helmeted Honeyeaters.

6.2.2 Spatial Correlation of Environmental Variation

Spatial correlation refers to the similarity of environmental fluctuations in different parts of the landscape—and, in the case of a metapopulation, in different populations. By "similarity," we mean the synchrony of these fluctuations rather than their magnitude. For example a habitat patch may receive much less rain than another, but some years both may receive above normal rainfall, and in other years both may receive below-normal rainfall. If the patches experience the same sequence of wetter-than-usual and drier-than-usual years, this means that the rainfall is spatially correlated, even though some parts of the landscape may get much more rain than others.

The importance of this factor can best be described by a simple example. Suppose that you need to evaluate the extinction risk of a metapopulation that consists of two populations. You have modeled each of these populations separately, and know that each has a 10% risk of becoming extinct in the next 100 years. You also know that these populations are in such different places that the environmental fluctuations they experience are not correlated. If their risks of extinction are mainly due to environmental fluctuations, we can assume that these risks are independent of each other. "Independent" means that if one population becomes extinct, the other may or may not become extinct in the same time step; in other words, extinction of one gives no information about the fate of the other.

Now we want to know the risk of extinction of the metapopulation, i.e., the risk that *both* populations become extinct within 100 years. To calculate this, we use a simple rule of probability, which says that when two events are independent, the joint probability that both will happen is the product of their constituent probabilities. Since each probability is 0.1, the joint probability is $0.1 \times 0.1 = 0.01$, so the risk that the metapopulation will become extinct in 100 years is 1%.

Now assume that the populations are in the same environment, and we know that if one becomes extinct, the other will as well. In other words, their dynamics are correlated, and their risks of extinction are fully dependent. In this case, the risk that both populations will become extinct is the same as the risk that one will become extinct (because they only become extinct together). So, the risk of extinction of the metapopulation is 10%, or 10 times higher than in the previous case of uncorrelated (independent) population fluctuations.

This aspect of metapopulation dynamics was first pointed out by den Boer (1968), who noted that when fluctuations were spread over a number of separate populations, the overall risk faced by the metapopulation was reduced. If the fluctuations in the environment are at least partially independent, so will be the fluctuations in population growth rates. Thus it will be less likely that all populations become extinct at the same time, compared to a case where the fluctuations are synchronous.

Correlation among the fluctuations of populations is often a function of the distance among them. If two populations are close to each other geographically, they will experience relatively similar environmental patterns, such as the same sequence of years with good and bad weather. This may result in a high correlation between the vital rates of the two populations. For example, Thomas (1991) found that Silver-studded Butterfly (*Plebejus argus*) populations that were geographically close tended to fluctuate in synchrony, whereas populations further apart (>600 m between midpoints) fluctuated independently of one another. Similarly, Baars and van Dijk

(1984) found that in two carabid beetles, *Pterostichus versicolor* and *Calathus melanocephalus*, the significance of rank correlation between fluctuations declined with increasing distance between sites.

When modeling metapopulations, the correlation among population fluctuations may be modeled as a function of the distance among habitat patches. This can be done by sampling the growth rates of each population from random distributions that are correlated, and the degree of correlation may be based on the distance among populations. (This is quite tedious to do manually, but very simple to do with a computer program; see Exercise 6.1.) This approach was used by LaHaye et al. (1994) to model correlated metapopulation dynamics of the California spotted owl (Figure 6.5; see the sample file OWL.MP). LaHaye et al. (1994) modeled this spotted owl metapopulation by making the growth rates of each population correlated with the growth rates of other populations. They calculated the degree of correlation based on the similarity of rainfall patterns among the habitat patches.

6.2.3 Dispersal Patterns

If extinct populations (i.e., empty patches) are recolonized by individuals dispersing from extant populations, a metapopulation may persist longer than each of its populations. Therefore, dispersal among local populations that leads to successful recolonization usually decreases extinction risk of the species.

In this chapter, we use the terms dispersal and migration interchangeably and define them as the movement of organisms from one population to another. Thus, migration does *not* mean back-and-forth seasonal movement between wintering and breeding locations.

The rate of dispersal is measured by the proportion of the individuals in one population that disperse to another. Suppose there are 100 individuals in population A; 5 of them disperse to population B, and 10 of them to population C; the rest stay in population A. In this case the dispersal rate from A to B is 5%, and from A to C it is 10%; and the total rate of dispersal from population A is 15%.

Dispersal rate depends, to a large extent, on species-specific characteristics such as the mode of seed dispersal, motility of individuals, ability and propensity of juveniles to disperse, etc. These factors will determine the speed and ease with which individuals search for and colonize empty habitat patches.

Dispersal rate between different populations of the same species may also differ a lot, depending on the characteristics of the particular metapopulation or of the specific population. For example, the habitat that separates two populations will affect the rate of dispersal between them.

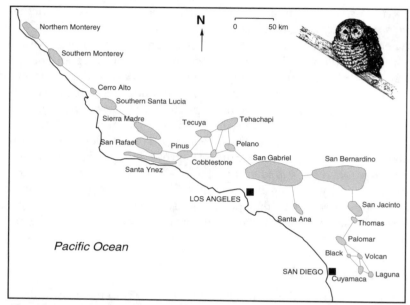

Figure 6.5. California Spotted Owl metapopulation (after LaHaye et al. 1994).

Two woodland patches with a connecting row of trees (a habitat corridor) between them may have a higher rate of dispersal than other populations separated by a highway (a barrier). Dispersal can also occur at different rates in two directions between two populations. For instance, individuals from local populations of a species along a river may migrate mostly or only downstream, but not upstream. In addition to these factors, human-mediated dispersal can have significant effects on extinction probabilities (we will discuss these later in this chapter). Below we discuss four other factors that affect dispersal rates, including the distance between populations, the abundance of individuals, their sex and age composition, and chance events.

6.2.3.1 Distance-dependent dispersal

Dispersing individuals may have a higher chance of ending up in a close patch rather than a distant patch. Thus dispersal may occur at a higher rate between populations that are geographically close. The relationship between dispersal rate and distance can be described as a declining curve. Such a function was used to model the dispersal of juvenile California Gnatcatchers (*Polioptila c. californica*), a threatened bird species (Figure 6.6).

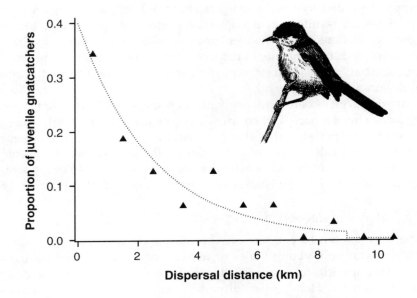

Figure 6.6. Proportion of dispersing California Gnatcatcher (*Polioptila c. californica*) juveniles as a function of distance (after Akçakaya and Atwood 1997).

The curve that summarized the dispersal-distance relationship is based on three parameters: average dispersal distance, maximum dispersal distance, and maximum dispersal rate (the y-intercept). In this case, the average distance traveled by dispersing juvenile gnatcatchers was about 2.5 km. This value determines how fast the curve declines as distance increases. The larger the average dispersal distance, the slower the curve declines. The maximum dispersal rate was 0.4, which is where the curve intersects the y-axis. In this figure, the maximum dispersal distance is set to 9 km, which is why at a distance of 9 km, the curve drops to 0 (no dispersal).

6.2.3.2 Density-dependent dispersal

In some species, the dependence of dispersal or dispersal rates on population abundance is an important aspect of the ecology of the species. For example, organisms may have a greater tendency to emigrate from their population under overcrowded conditions, resulting in not only a larger number, but also *a greater proportion* of individuals leaving the population as density increases. This causes the dispersal rate to be an increasing function of abundance. A similar effect can also occur in plant populations if, for example, high density causes an aggregation of frugivorous organisms that help dispersal, resulting in a higher proportion of seeds dispersed from larger populations than smaller ones. Under this model of density-dependent dispersal, the rate of dispersal (and consequently the probability of recolonization) is an increasing function of the density of the source population.

Another type of density dependence in dispersal rates is called the stepping stone effect. It occurs when smaller populations are used only as a short stop during dispersal, rather than for settling. In this case, the organisms have a higher tendency to emigrate from *smaller* populations. For example, the dispersal rates of voles from smaller islands were found to be higher than dispersal rates from larger islands in an archipelago in Finland (Pokki 1981).

6.2.3.3 *Age- and stage-specific dispersal*

Dispersal rates can be age- or stage-specific, such as when only immature individuals or only young males or females disperse to other habitats. In most plant species, for example, dispersal occurs only in the seed stage. Female Helmeted Honeyeaters disperse from their natal colony to breed. And in most territorial bird species (such as the Spotted Owl and the California Gnatcatcher), juveniles disperse farther away than adults. If the age and stage structure of the subpopulations (i.e., the proportions of individuals in different stages) are different from each other, this factor may have a significant effect on metapopulation dynamics.

6.2.3.4 *Stochasticity*

The proportion of individuals migrating from one population to the other may also change in a random way. Similar to demographic stochasticity in survival and reproduction discussed in previous chapters, the fact that only whole numbers of individuals can migrate from a given population to another will introduce variation. Because of this similarity, if you specify demographic stochasticity in RAMAS EcoLab, the program will sample the number of migrants from a binomial distribution.

6.2.4 Interaction Between Dispersal and Correlation

The risk of extinction or decline of a species is determined by the factors discussed above and by the interrelationships among them. On the one hand, the extinction probabilities of local populations that are relatively far from each other will be largely independent; hence perhaps the metapopulation will have a lower overall extinction risk. On the other hand, dispersal rates (and hence recolonization chances) in such a metapopulation will probably be lower compared to a metapopulation with closer local populations. Thus there is always a trade-off between similar environments (and, consequently, correlated extinctions) and higher dispersal rates for close populations.

Another important aspect of spatial structure is the *interaction* between these two factors. Interaction refers to the changes in the effect of one factor that depend on another factor. In this case, how much dispersal helps reduce the extinction risk of the species will depend on the similarity of the environments that the populations experience. Consider the extreme case of perfect correlation of environments. In such a case, populations will almost always become extinct at the same time period, and whether there is dispersal or not before this time will not change the extinction risk. Dispersal will decrease this risk only if the populations become extinct at different times so that the extinct patches have a chance of being recolonized by migrants from extant populations.

The interaction between dispersal and correlation is demonstrated in the results of a study that compared the extinction risk of a single population, with that of a metapopulation that consisted of three small populations (Figure 6.7). The three populations had the same total initial abundance as the single large population, and their risk of extinction in the next 500 years was lower than that of the large population, when correlation was low (the left end of the curves). In addition, when the correlation was low, higher rates of dispersal caused lower extinction risks. When correlation was high, the single large population had a lower risk, and the effect of dispersal rate was negligible (the four curves get closer towards the right end of the graph). See Exercise 6.1 for another demonstration of the interaction between correlation and dispersal.

6.2.5 Assumptions of Metapopulation Models

One of the first models developed for metapopulation dynamics was by Levins (1970). In this model, the proportion of occupied patches (p) is determined by colonization of empty patches and the extinction of occupied patches:

$$dp/dt = m\,p\,(1-p) - E\,p$$

In this differential equation, dp/dt refers to the rate of change in the proportion of occupied patches. The colonization parameter (m) is defined as the probability of successful migration from an occupied patch to any other patch per unit time. The parameter E is the probability of extinction of a

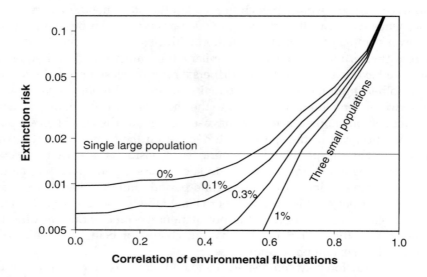

Figure 6.7. Extinction of a single large population (horizontal line), and three small populations (curves) as a function of the correlation of environmental fluctuations. Each curve represents a different simulation of the three-population model with different dispersal rates (0% to 1%). After Akçakaya and Ginzburg (1991).

given local population in a unit time interval. Colonization (the first term in the equation) is assumed to be proportional to the product of occupied patches p and unoccupied patches $1-p$, and extinction (the second, negative term, Ep) is proportional to the number of occupied patches. This model predicts that the proportion of occupied patches will reach the equilibrium number,

$$p^* = 1 - E/m$$

In other words, the species will persist (or, the proportion of occupied patches will be greater than zero) if the rate of colonization (m) exceeds the rate of extinction (E).

This model predicts the future of the population in terms of the number of occupied patches, instead of the number of individuals. As a result, it does not incorporate the within-population factors (such as density dependence) that we have discussed in previous chapters. Instead, it assumes patches are either fully occupied (the population is at the carrying capacity) or they are empty (population is extinct). It also assumes that all patches are equal in terms of risk of extinction and chance of being colonized. Further, it assumes that extinction of each population is independent of others.

Other metapopulation models that were based on Levin's model make different assumptions regarding extinction and colonization rates. However, most make the basic assumption that patches are either occupied or empty. For this reason, we call these occupancy models.

The models we will develop in the exercises using RAMAS EcoLab make a different set of assumptions. As in previous chapters, they describe each population by its abundance, and they incorporate environmental and demographic stochasticity. Their two basic assumptions are: (1) there is no age, stage, sex, or genetic structure (assumption 4 in Chapter 1), and (2) the dynamics can be approximated by pulses of reproduction and mortality; in other words, they happen in discrete time steps (assumption 6 in Chapter 1). In terms of metapopulation-level factors, they assume that the correlation of environmental fluctuations and the rate of dispersal among populations can be described as a function of the distance among the populations. The distances among populations are calculated by the program based on the coordinates of each population. Thus, in these models, the location of each population makes a difference in terms of the dynamics of the metapopulation.

6.3 Applications

The existence of a species in a metapopulation necessitates a different approach than that used for single populations. This is especially true in the case of applications in population ecology, such as assessing human impact on threatened and endangered species, and evaluating options for management and conservation. Each population of a metapopulation may be impacted by human actions to a different degree. These effects may be observed in terms of lower vital rates or carrying capacities. In addition, a metapopulation can be impacted by human actions in ways that are not easily noticeable with a single-population approach. For example, logging can increase the fragmentation of old-growth forests; agricultural expansion can change the spatial distribution of remaining native habitat; highways and power lines can decrease dispersal among populations and lead to increased isolation of the remaining patches. Below, we will discuss types of management options and impact assessments that are relevant for metapopulations.

6.3.1 Reintroduction and Translocation

The existence of multiple populations also brings with it new types of management options that do not exist for single populations. These include reintroduction and translocation.

The IUCN (1987) defines reintroductions as the "intentional movement of an organism into part of its native range from which it has disappeared or become extirpated as a result of human activities or natural catastrophe." The intention is to establish a self-maintaining, viable population in an area that was previously inhabited by the same species. Reintroductions should be distinguished from translocations, the movement of individuals from one patch of habitat to another.

Reintroductions are a risky strategy. A provisional survey noted 220 plant reintroduction projects conducted world-wide between 1980 and 1990, involving 29 different plant families (WCMC 1992, Chapter 34). Preliminary indications are that many have been unsuccessful because of poor horticultural practice, poor ecological understanding, lack of post-planting maintenance and monitoring. Griffith et al. (1989) reviewed translocations and reintroductions around the world, and documented more than 700 cases per year, mostly in the United States and Canada. They found that translocation of game species constituted 90% of translocations and that they had a success rate of 86%. Translocations of threatened species made up the remainder, and their success rate was just 44%. Dodd and Siegel (1991) found that only 19% of translocations involving reptiles and amphibians were successful.

There are many reasons for failure, including the quality of habitat into which the animals are released and whether the individuals are wild or captive bred. Social structure of the population may interfere with the success of the captive bred individuals reintroduced to a wild population (e.g., see Akçakaya 1990). The design of the reintroduction program is also important, and includes such factors as the number, sex, age composition, and social structure of the released population, the provision of supplementary food, and the use of a single release or multiple releases over many years. To succeed, reintroductions need to be carefully planned, executed, and monitored.

Translocations may augment the natural dispersal rate and can be effective since they can easily be planned to be density-dependent, by moving individuals from high-density populations to empty or low-density patches (see Section 6.2.3.2 on density-dependent dispersal). Whether translocation will increase the persistence of the metapopulation depends on many factors; they include both the spatial factors we discussed in this chapter and others such as the risk of injury or mortality due to handling and transport by humans.

Planning of reintroductions and translocations often requires a metapopulation approach. If a species is to be reintroduced, questions such as "Is it better to reintroduce 100 animals in one patch, or 50 in each of two patches?" require an understanding of the metapopulation dynamics of the species. The design of the optimal translocation schedule involves questions such as "How many?" and "From which population, to which population?", suggesting a metapopulation approach.

6.3.2 Corridors and Reserve Design

Dispersal between populations can also be increased by building or protecting habitat corridors, which are linear strips of habitat that connect larger patches of habitat. Although corridors for dispersal have long been recommended as a conservation measure, there is often little reliable data on how often corridors are used by a given species, and how much they increase the persistence of the species (Simberloff et al. 1992). Corridors may increase dispersal rates, but not in all cases; and while increased dispersal usually decreases extinction risks, models suggest that in some cases the reverse may happen. Increased dispersal from source populations to sink populations might increase extinction risks (Akçakaya and Baur 1996). Corridors may also help the spread of catastrophes such as fires and disease epidemics, and act as sink populations (because of strong edge effects). In summary, it is not possible to make a general statement about the effectiveness of corridors as a conservation measure; this depends on the particular metapopulation, and the particular landscape it lives in.

Another conservation option that relates to metapopulation dynamics is the design of nature reserves, which involves selection of habitat patches that will give the most protection to the species in question. This provides a practical example of the interactions and trade-offs among components of spatial structure. One question that occupied conservation biologists for many years was whether a single large or several small reserves of the same total area will provide better protection for a species against extinction (known by the acronym SLOSS, single large or several small). On the one hand, several small populations may have a lower extinction risk if the rate of dispersal is high enough and the degree of spatial correlation of environments is low enough. This is because a single large population will not benefit from uncorrelated environmental fluctuations; if it becomes extinct, it cannot be recolonized. On the other hand, compared to a large population, each of the small populations will be more vulnerable to extinction due to environmental and demographic stochasticity. Thus if they become extinct at the same time, or if the extinct ones cannot be recolonized from others, a metapopulation of several small populations may have a higher extinction risk than a single large population (see Section 6.2.4).

Thus there is no general answer to the SLOSS question. The answer depends not only on these two factors (degree of correlation and chances for recolonization), but also on other aspects of metapopulation dynamics, such as the configuration, size and number of populations, their rates of growth, density dependence, carrying capacities, etc. However, metapopulation modeling allows us to find an answer to the SLOSS question for specific cases, and to evaluate different configurations of habitat patches selected as nature reserves. Actually, conservation biologists are rarely faced with the question in such simple terms. Often the monetary or political cost of acquiring a patch for a reserve might not be related to its size; in other cases the size (or even the carrying capacity) of a patch might not be directly related to its value in terms of the protection it offers. A small patch that supports a stable population might contribute more to the persistence of the species than a large patch that is subject to greater environmental variation or human disturbances. Thus it is much more productive to evaluate reserve design options for a specific case, using as much of the available empirical information as possible, than trying to find generalities that may or may not apply to specific cases.

6.3.3 Impact Assessment: Fragmentation

It is important to note that the trade-off between large and small reserves we discussed above only applies to the case where the total area of a single large reserve is roughly equivalent to the total area of small reserves. If the metapopulation of several small populations in the above example has formed from a single large one as a result of habitat fragmentation, the answer to the above question will be much less ambiguous. A fragmented habitat that has several small patches certainly contains a smaller (and a more extinction-prone) total population compared to the original nonfragmented habitat. The reason is that, as a result of fragmentation, (1) the total area of habitat is reduced; (2) the movement of individuals (migration, dispersal) is restricted; (3) the resulting habitat fragments are generally no more independent of each other than they were before fragmentation; (4) populations in fragmented habitats may have lower vital rates because of edge effects (see Section 6.1.2). Each of these point to an increased risk of extinction due to fragmentation, but they may also have exceptions. For example, while it is usually true that parts of the habitat would not have more independent (uncorrelated) environmental fluctuations after fragmentation, certain factors, such as slower spread of fires or diseases may make the habitat fragments more independent than parts of a single large habitat.

Another factor to consider in applying metapopulation concepts to the assessment of fragmentation effects (and reserve design decisions) is the change in patch sizes over time. In the above discussion about SLOSS, we assumed that the size of the patches that are selected remains the same, or at least, the change in size does not depend on whether one large or several small patches have been selected. In fact, because of edge effects, the original habitat in smaller patches may decrease in size faster than in larger patches. (This effect can be simulated in RAMAS EcoLab by specifying a negative rate for the **Temporal trend in K** parameter in **Populations**; see below.)

All these interactions and complexities make it impossible to assess species extinction risks or address questions about conservation and management based solely on intuition or simple rules of thumb. When conservation decisions are based on rules of thumb, a number of assumptions are made. Building models forces us to make these assumptions explicit, and whenever data are available, to replace these assumptions with model parameters estimated for the specific case at hand.

6.4 Exercises

In these exercises, we will use the "Multiple Populations" component of RAMAS EcoLab. The following paragraphs describe the special features of this component.

An Overview of the Program

The metapopulation models you can build with this program are based on the unstructured models (models with no age or stage structure) we worked on in Chapters 1, 2, and 3. In other words, each population is modeled in the same way we modeled single populations using the "Population growth (single population models)" program. The first window under the Model menu (**General information**) is the same as in that program; here you can specify a title, comments, the number of replications and time steps, and whether to use demographic stochasticity. The second window under the Model menu (**Populations**) is also similar to the ones you used in Chapters 2 and 3, but with two differences:

(1) On the left side of the window, there is a list of the populations in the metapopulation. On the right side, there are the parameters for the population that is highlighted on the list. There is one such set of parameters for each population. So, to edit the parameters of a population, first click on its name on the list, then click on the appropriate parameter and type in the new value.

(2) There are several additional parameters for each population. These include the name and coordinates of the population, so that each population comprising the metapopulation can be distinguished from others. Another new parameter is "Temporal trend in K," which describes how the carrying capacity of the population changes through time.

The "Display" button in **Populations** lets you view four graphs related to the parameters of the population (the one that is highlighted on the list). The first two graphs are related to density dependence and are the same as in previous components. The third shows a histogram of the rates of dispersal from the specified population to each of the other populations (see below). The fourth graph shows the change in carrying capacity through time. The "Carrying capacity (K)" parameter determines where the line in this graph starts from (the y-intercept), and the "Temporal trend in K" parameter determines how much it increases or decreases per time step (the slope of the line).

Another feature of this program is the **Metapopulation map** displayed in the main window. The relative position of each population is calculated from the X and Y coordinates specified in **Populations** screen. The areas of population circles are proportional to their carrying capacities if there is density dependence; otherwise the initial abundance is used. The brightness (or "fullness") of the population circles shows the relative degree of occupancy, calculated as the ratio of initial abundance to carrying capacity.

If there is dispersal from one population to another, this is indicated by a line between the two population circles. This line is *not* drawn if the expected number of migrants (dispersal rate multiplied by the carrying capacity of the source population) is less than 0.1. The line connects to the source population, but, at its other end, there is small gap (not visible in some cases) between the end of the line and the target population, so that you may understand the direction of dispersal. If there is dispersal in both directions, then the two lines will be superimposed, and there will not be any gaps at either end.

Also drawn on the screen are various geographic features from an ASCII file (which is specified as the "Map features" parameter in **General information**). The format of this file follows that for RAMAS Metapop and RAMAS GIS (Akçakaya 1998), but for these exercises, you don't need to know the format.

There is also an additional window that is not in the single population program. The **Dispersal and Correlation** window can be used to specify the rate of dispersal between populations and the similarity of their fluctuations, based on the distance between them. The distances are calculated from the x-

and *y*-coordinates specified in **Populations**. Dispersal rates are specified as a function of three parameters: average dispersal distance, maximum dispersal distance, and maximum dispersal rate (see Section 6.2.3.1).

Correlations are also specified as a similar function of distance. To keep things simple, only one parameter needs to be specified: the correlation at a distance of 100 units (a unit depends on the units used to specify the coordinates of populations). If this parameter is 1.0, all populations are fully correlated; if it is 0.0, all populations are uncorrelated (have fully independent fluctuations). If it is 0.5, then fluctuations of two populations separated by 100 units (say, km) will have a correlation coefficient of 0.5. Populations closer than 100 km will have a higher correlation, and those farther apart will have a lower correlation. (For the curious: the shape of both functions is that of the negative exponential function $y = a \cdot \exp(-d/b)$, where d is the distance and a and b are parameters. The dispersal function is truncated at the maximum dispersal distance).

In **Dispersal and Correlation**, click the "View" buttons to view two graphs that show dispersal rate and correlation as functions of the distance between two populations. You can also see the rates of dispersal from one population to all other populations. For this, go to **Populations**, click "Display" and select "Dispersal rates." The horizontal line shows the total rate of dispersal from this population (which is also displayed numerically at the top of the screen), and the vertical bars show the proportion of individuals dispersing to other populations.

Exercise 6.1: Spatial Factors and Extinction Risks

This exercise demonstrates the effect of three spatial factors on metapopulation extinction risks: number of populations, correlation among population growth rates, and rate of dispersal among populations. The exercise consists of several models of hypothetical metapopulations. These models are saved in files that are described below. The files may also contain results. If so, you do not need to run the simulation. If a file does not contain results, the program will tell you this when you attempt to view a result screen, and you will need to run a simulation with each model, and save the results.

Step 1. Single large population

We begin with a single population which has a carrying capacity of 100 individuals. Open the file 1LARGE.MP, which contains such a population model. To review the specifics of the model, you can go through the input parameters in **General information** and **Populations**. Other input windows do not contain any information since there is only one population. Select "Trajectory summary" from the Results menu to view the plot of population size though time.

For the estimate of extinction risk, select "Extinction/Decline" from the Results menu. The graph shows the probability that the population will fall (during the next 30 time periods) below each threshold level in the x-axis. From the graph, read the risk of falling below 5 individuals, and record this number.

Step 2. Correlated environments; no dispersal

The file 5SM-C-I.MP contains a metapopulation model of five populations, each with a carrying capacity of 20 individuals. Thus the total size of this metapopulation is the same as that of the single large population in the previous example. The other population parameters (such as growth rate, its standard deviation, survival rate, the duration of the simulation, etc.) are the same as those of the single population.

Open the file 5SM-C-I.MP (the filename summarizes "5 small; correlated; isolated"). Examine the input parameters in each of the three windows under the Model menu. Notice the following:

(1) In **General information** and **Populations**, all parameters except the initial population size (N_0) and carrying capacity (K) are the same as in the single large population; N_0 and K for each of the five populations are one-fifth of their values for the single large population.

(2) In **Dispersal and Correlation**, the maximum dispersal rate is 0, and correlation at d=100 is 1. The first parameter means that there is no dispersal among the populations, and the second means that environmental fluctuations of the populations are fully correlated. Click the two "View" buttons to see the two functions: the dispersal function is zero for all distances, and the correlation function is one.

(3) There are no lines among the populations in the **Metapopulation map** displayed in the main window (because there is no dispersal).

Again view the Extinction/Decline risk (do not change the scales). If you've changed any of the input parameters, the program will give a warning; in this case load the file again. You will notice that this metapopulation has a higher extinction risk than the single large population, and the risk curve is above the risk curve of 1LARGE.MP. Again, record the risk of falling below 5 individuals.

As you have noticed from your inspection of the input screens, the growth rates of the five populations are correlated, and there is no dispersal among the populations. Hence this metapopulation represents the worst combination of these two factors: the populations will become extinct at about the same time, and there will be no chance of recolonization. This results in a much higher extinction risk compared to the single large population, even though the total population sizes are the same.

To see a visual demonstration of the effect of correlation, start the simulation by selecting **Run** from the Simulation menu. The program will show the spatial structure of the metapopulation, updated after every time step. If the map is changing too fast, increment the counter on the top of the window to change the simulation delay.

At each time step, the abundance of the five populations is used to update the map. The shading in each population represents how "full" the patch is, i.e., how close the population is to the carrying capacity of the patch. Notice that when one patch is completely full, the others are more likely have a large population than be extinct. This is because of the high correlation among population growth rates. However, this is not always the case; you may notice that sometimes a population becomes extinct while others are still extant. This is because demographic stochasticity introduces variation that is independent for each population.

You can click the left-most button on top of the window to turn off the map and complete the simulation faster; or you can press [Esc] to terminate the simulation (the program will stop running the simulation after the current replication is completed). Close the simulation window.

If you turn off demographic stochasticity (by clearing its box in **General information**), the population sizes will become fully correlated. However, turning off demographic stochasticity also decreases the risk of extinction, so you will notice fewer "empty" patches during the simulation than you did with demographic stochasticity turned on.

Step 3. Independent environments; no dispersal

The file 5SM-U-I.MP contains a similar model, but the populations are independent: their growth rates have zero correlations. (The filename summarizes "5 small; uncorrelated; isolated"). You can check this by opening this file and selecting **Dispersal and Correlation** under the Model menu. The correlation coefficient (at $d = 100$) is zero. Click the "View" buttons to see the dispersal and correlation graphs. Other parameters of this model are the same as in the previous file.

View the extinction risk curve, and record the risk of falling below 5 individuals. The lower extinction risk is a result of the independence of environmental fluctuations, which cause fewer of the populations to become extinct at the same time than with the correlated fluctuations in the previous model.

Now run the simulation to observe the effect of independent fluctuations. You will notice that the sizes of the five populations (represented by the density of shading in the population circle) change independently; when one or two populations have high densities, others may have medium or low densities, or even be extinct. Notice that there are no lines (that represent dispersal) among the five populations. Thus the decrease in extinction rate occurred even though the extinct patches did not have a chance to be recolonized from extant populations. (You can terminate the simulation by pressing (Esc), and return to the main menu by pressing any key after the end of the simulation.)

Step 4. Dispersal; uncorrelated environments

To see the effect of dispersal in decreasing extinction risks, open the file 5SM-U-D.MP. This file contains a model of the same metapopulation as the previous model, but with a moderate rate of dispersal among the populations. (The filename summarizes "5 small; uncorrelated; with dispersal"). Open this file, and select **Dispersal and Correlation**. Note that the maximum dispersal rate (the y-intercept) is now 1.0. Click the "View" button for dispersal to see the dispersal-distance function, which gives the proportion of dispersers as a function of the distance between the source and target populations. The distances among the populations in this model range from 10 to 20 (which you can calculate from the coordinates). The dispersal-distance function gives dispersal rates ranging from 0.007 to 0.082 between any two populations in this model. If you added the rates of dispersal from one population to all others, you would get about 14 to 18% of the individuals in each population dispersing to the other four populations.

You can also see the rate of dispersal from one population to each other population. For this, go to **Populations** screen and select a specific (source) population (by clicking on the name of the population in the list on the left). Then click "Display," and select "Dispersal". The horizontal line shows the total rate of dispersal from this population (which is also displayed numerically at the top of the screen), and the vertical bars show the proportion of individuals dispersing to each other population.

View the Extinction/Decline risk (without changing the scales), and record the risk of falling below 5 individuals. When you now run the simulation a line will appear between two populations if, at that time period, there is dispersal between those two populations (in either direction). At some time steps, you may not see a line between some populations; this is because if the size of a population gets very small, the number of emigrants from that population may be rounded-off to zero. If, for example, dispersal rate from one population to another is 0.05 and the number of individuals in the source population at a particular time step is 9, the number of migrants for that time step will be $0.05 \times 9 = 0.45$, and will be truncated to zero. In addition, when demographic stochasticity is turned on, the number of migrants is sampled from a binomial distribution (see the section on "Stochasticity" above).

Step 5. Dispersal; correlated environments

The last file, 5SM-C-D.MP, contains a model of the same metapopulation as the previous model, but in a correlated environment (the filename summarizes "5 small; correlated; with dispersal"). View the extinction risk result, and record the risk of falling below 5 individuals.

Step 6. Summarizing results

Combine all the risk results you have recorded into the table below. Compare the risks of falling below 5 individuals with the five models. Does a single large population have a lower or higher risk than several small populations?

Risk of falling below 5 individuals
with the single population model: _____

Metapopulation models:	Risk with no dispersal	Risk with dispersal	Reduction in risk due to dispersal
Full correlation	5SM-C-I.MP	5SM-C-D.MP	
No correlation	5SM-U-I.MP	5SM-U-D.MP	

Step 7. Effect of dispersal in reducing extinction risks

To see how dispersal effects extinction risks, subtract the risk in models with dispersal from the risk in models without dispersal. There are two models without dispersal and two with dispersal, so you calculate the effect of dispersal in reducing risks in two different ways. One of these assumes full correlation, and the other assumes no correlation. Compare the effectiveness of dispersal in preventing extinctions under these two assumptions. Explain the difference (see Section 6.2.4).

Exercise 6.2: Habitat Loss

As we mentioned in Section 6.3.3, an important factor in applying the metapopulation concepts to conservation decisions is the change in patch sizes over time. With RAMAS EcoLab you can simulate the change in the carrying capacity of a population if the population has density-dependent population growth.

Step 1. To observe this effect, first load 1LARGE.MP, the single population model, and select the Extinction/Decline screen from the Results menu.

Step 2. Select **Populations** from the Model menu, and change the parameter **Temporal trend in K** to −3. Click "Apply," then "Display" and select **Habitat change**. This graph shows the change in the carrying capacity over the duration of the simulation. The carrying capacity is decreasing by 3 at each time step. Close the window and click "Cancel" (twice). Run a simulation. Notice that the circle representing the population gets smaller at each time step. The size of the circle is proportional to the carrying capacity.

Step 3. If you want to complete the simulation faster, click the text button (fist button from left on top of the window) to turn off the map. After the end of the simulation close the simulation window, and select **Extinction/Decline** under the Results menu. How did the risk curve change? What is the risk of falling below 5 individuals? How does this risk compare with the risk of falling below 5 individuals when the model did not include any habitat loss.

Step 4. Select **Trajectory summary** to see the predicted decline in the population size over the next 30 years. You can run a similar simulation with one of the metapopulation models.

The effect of habitat loss will depend, among other things, on the number of populations for which you specified a negative change in carrying capacity, and at the rate of loss specified with the value of the **Temporal trend in K** parameter. This parameter can also have a positive value representing an increase in habitat. To see a model with both increasing and decreasing patches, load METAPOP6.MP, and run a simulation.

Exercise 6.3: Designing Reserves for the Spotted Owl

This exercise concerns the metapopulation dynamics of the California spotted owl (see Section 6.2.2). The model is based on one of the models used by LaHaye et al. (1994) to explore the effects of spatial factors on this metapopulation.

Step 1. Start RAMAS EcoLab, select "Multiple population models," and load the file OWL.MP. This file contains a metapopulation model of the California Spotted Owl discussed above.

Step 2. The model predicts a fast decline of this metapopulation. For this exercise, we will assume that the reason for the decline is a decrease in habitat quality in recent years. We will also assume that it is possible to improve the habitat quality but that it costs a lot of money. How much it costs depends on the size of the habitat to be improved. As you see on the map, each population has a different size. The cost of habitat improvement for a population is $1,000 multiplied by the carrying capacity (K) of the population. For example, improving the habitat in the first population (N. Monterey) would cost $100,000. Your job is to decide in which populations to improve the habitat. You have a total of $500,000 to spend. Habitat improvement in any patch results in an increase in the growth rate from 0.827 to 1.01.

Step 3. Select **General information**, set the number of years to 20, and number of replications to 1,000 (if your computer is very slow, you can use a smaller number such as 300 or 500), and run a simulation. The simulation may run faster if you turn off the display of the map. Save the file with a new name.

Step 4. Then select as many populations as you can with your $500,000, and increase the growth rates of the selected populations to 1.01. Keep the other populations in the model, and leave their growth rates as they are. The sum of the carrying capacities of the populations you selected should not exceed 500. Then save the file under a different name, and run a simulation. After the simulation is over, save the file again; investigate and record the results.

Step 5. Your plan of habitat improvement will probably not prevent the decline of the metapopulation, but it can slow it down so as to gain some more time for further conservation actions (or perhaps for raising more money to improve more habitat). So, your criteria for success should not be whether the average size of the metapopulation is increasing or decreasing. Instead, find a probabilistic measure of whether your plan is successful or not. For example, you can use the risk of falling below 100 individuals as your measure. Compare this result for the two simulations (in steps 3 and 4 above). How much did the result improve?

Step 6. Now select different combinations of populations to improve (again, the sum of the carrying capacities of the populations with growth rate equal to 1.01 cannot exceed 500). Run more simulations and compare the results.

Step 7. Describe which populations you selected in each case and how you made your selection of populations. Random selection is acceptable. Other criteria may include: few largest populations, many small populations, most geographically spread subset of populations, all populations in the north, or south or the center, all populations away from the shore, etc.

Step 8. Describe which selection was the most successful. If two or more selections give similar risks of falling below 100 owls, compare them with respect to the risk of falling below 200 owls.

NOTES:

Given the parameters and assumptions of the model, there is probably a single best solution to this exercise. However, the number of combinations is quite large, and you are not expected to try all possible subsets. If this were a real case, then you'd probably want to cover all possible options before you spend your half-a-million dollars.

When you find out that a particular combination gives much better results than others, do not be tempted to generalize. The point of this exercise is that when populations are distributed in space, where they are makes a difference. But the rule of selection that is best will probably be different in each case.

6.5 Further reading

Askins, R. A. 1995. Hostile landscapes and the decline of migratory songbirds. *Science* 267:1956–1957.

Gilpin, M. E. 1987. Spatial structure and population vulnerability. In *Viable Populations for Conservation*, M.E. Soulé (Ed.), pp 126–139. Cambridge University Press.

Hanski, I. 1989. Metapopulation dynamics: does it help to have more of the same? *Trends in Ecology and Evolution* 4:113–114.

Harrison, S. 1991. Local extinction in a metapopulation context: an empirical evaluation. *Biological Journal of the Linnean Society* 42:73–88.

Simberloff, D., J. A. Farr, J. Cox, D. W. Mehlman. 1992. Movement corridors: conservation bargains or poor investments? *Conservation Biology* 6:493–504.

Chapter 7
Population Viability Analysis

7.1 Introduction

So far, we have dealt with topics that are general in population ecology. In this and the next chapter, we will concentrate on two specific areas where the principles of population ecology may be applied. One specific area to apply the methods and concepts discussed in previous chapters is population viability analysis (frequently abbreviated to PVA). Population viability analysis is a process of identifying the threats faced by a species and evaluating the likelihood that the species will persist for a given time into the future (Shaffer 1981, 1987, 1990; Gilpin and Soulé 1986; Boyce 1992). The process of PVA is closely related to determining the minimum viable population (MVP), which is defined as the minimum number of individuals that ensures a population's persistence. As we have demonstrated in previous chapters and will discuss below, the size of a population is only one of the characteristics that determine the chances of persistence; thus PVA can be thought of as a generalization of the MVP concept.

Population viability analysis is often oriented towards the management of rare and threatened species; by applying the principles of population ecology, PVA seeks to improve the species' chances of survival. Threatened species management has two broad objectives. The short-term objective is to minimize the risk of extinction. The longer-term objective is to promote conditions in which species retain their potential for evolutionary change without intensive management. Within this context, PVA may be used to address three aspects of threatened species management (Possingham et al. 1993):

(1) *Planning research and data collection.* PVA may reveal that population viability is insensitive to particular parameters. Research may be guided by targeting factors that may have an important effect on extinction probabilities.
(2) *Assessing vulnerability.* PVA may be used to estimate the relative vulnerability of species to extinction. Together with cultural priorities, economic imperatives, and taxonomic uniqueness, these results may be used to set policies and priorities for allocating scarce conservation resources.
(3) *Ranking management options.* PVA may be used to predict the likely responses of species to reintroduction, captive breeding, prescribed burning, weed control, habitat rehabilitation, or different designs for nature reserves or corridor networks.

You have already applied principles and methods of population ecology to these aspects of PVA. For example, in Exercise 2.4, you analyzed the sensitivity of Muskox population viability to the parameters of a simple model. In Exercise 5.3, you compared the effects of two conservation measures on the viability of sea turtles, and in Exercise 6.3, you tried to find the reserve design option that maximizes the viability of a California spotted owl metapopulation. In this chapter, we will discuss population viability analysis in more detail, beginning with a review of extinctions.

7.2 Extinction

Population viability analysis deals with one aspect of population dynamics, namely the decline and extinction of populations. It is therefore relevant to review briefly what we know about extinction so that we understand the motivation behind PVA's methods and concepts.

7.2.1 Extinction in Geological Time

More than 99% of species that have ever existed are now extinct, and most species have a lifetime of around 1 to 12 million years. Many of the species that existed in the past were not eliminated in the sense that we think of extinction today. Rather, natural selection and mutation have essentially transformed many species, a process known as phyletic evolution (or evolution within a single lineage, without speciation). In addition, there have been a number of events in the geological past in which substantial proportions of the biota then existing were lost. These are termed mass extinction events. The most severe extinction event for marine families occurred 245 mya (million years ago) during the late Permian period, at which time more than half of the families of marine animals and tetrapods and nearly half of the number of fish families were lost. About 95% of all species were lost. The late Permian event is generally accepted to have occurred over a period of 5–8 million years, and appears to have been associated with global physical changes including climate change and volcanic activity.

The most recent mass extinction, which marks the boundary between the Cretaceous and Tertiary periods 65 mya, is the best documented. There is some evidence that it was associated with an extra-terrestrial impact although the cause remains controversial. The late Cretaceous event resulted in a decline of about 15% of marine families and 40% of tetrapod families in the fossil record. The loss of plant species was unusually high compared to other mass extinction events. More than 70% of all species were lost.

Another interesting type of information from the fossil record concerns length of time each species existed. The lifetimes of species (i.e., times to extinction after the initial radiation) are usually distributed asymmetrically; they are strongly skewed to the right (Figure 7.1) with most species persisting for time periods less than the average within any one taxon, and a few species persisting for much longer times.

In this kind of statistical distribution, the median time to extinction (the time at which half the species within a taxon have become extinct) is less than the average time to extinction. In the fossil record, the average rate of loss of species globally (both by species extinction in the usual sense of the word, and by phyletic evolution) has been of the order of 1 species per year. This estimate was first made by Charles Lyell and several more recent estimates have resulted in approximately the same number.

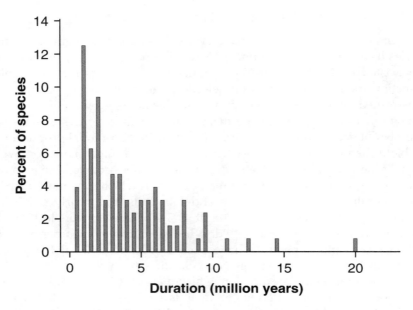

Figure 7.1. The time to extinction of species in planktonic foraminifera during the Paleogene radiation (after Levinton and Ginzburg 1984).

7.2.2 Current Extinction Rates

Most recorded extinctions over the last few hundred years are from mammals, birds, and terrestrial snails because the taxonomy of these groups is relatively complete. For most other vertebrates and almost all other invertebrates there is no information on extinction rates in recent history. The main difficulty is that most taxa have not been described or named. The vast majority of even the described species are not monitored, and species may be locally or globally eliminated without our knowledge of the event. Only named species are recorded as extinct. Species are presumed to be extinct when a specific search has not located them, when they have not been recorded for several decades, or when expert opinion suggests they have been eliminated. These rules strongly suggest that our knowledge of extinction rates, even in relatively well known groups, will tend to underestimate the true rate. Nevertheless, there have been a total of about 490 animal extinctions recorded globally since 1600 (WCMC 1992; see Figure 7.2 and Table 7.1).

There has been a preponderance of extinctions on islands compared to rates on continents. For example, 75% of recorded animal extinctions since 1600 (about 370 extinctions out of 490) have been of species inhabiting islands, even though islands support a small fraction of the number of

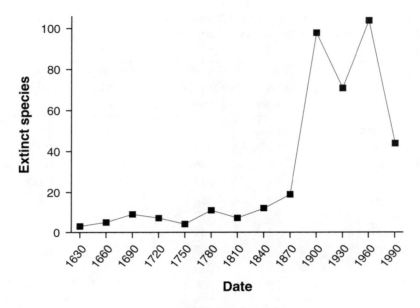

Figure 7.2. The number of recorded animal extinctions in recent history (after Jenkins 1992). The values represent the number of extinctions recorded within the 30-year period up to the date labeling the class.

animal species found on continents. The apparent decline in extinctions in the last 30 years to 1990 (Figure 7.2) is at least partly due to the fact that there is a time lag between extinction events and their detection and recording. A number of species are likely to have become extinct recently without yet having been recorded as such. It is also possible that conservation efforts over the last 30 years have slowed the rate of extinctions. Attention usually is focused on high-profile species (such as mammals and birds) and many recent recovery efforts have been successful, at least in the short term.

There is little doubt that current extinction rates are considerably higher than those observed globally over the last 50 million years. Some simple calculations serve to illustrate the point. Assume an average life-span for a species of about 4 million years and a total of about 10 million species. If an average species lives for 4 million years, we may assume it has an average probability of extinction of 1/4,000,000 per year. Multiplying this number with the total number of species (10 million) gives the rate of 2.5 species per year. Thus, we should expect an average of between 2 and 3 extinction events per year. Various people have estimated the background rate of extinction in geological time to be between 1 and 5 species globally per year,

Table 7.1. Species in major plant and animal taxa that are known to have become extinct since 1600 or are currently listed as threatened by IUCN.

Taxon	No. of extinct species	No. of threatened species	Total number of recorded species (x1000)	% extinct
All animals	485	3565	1400	0.04
Molluscs	191	354	100	0.2
Crustaceans	4	126	40	0.01
Insects	61	873	1000	0.006
Vertebrates	229	2212	47	0.5
Fish	29	452	24	0.1
Amphibians	2	59	3	0.1
Reptiles	23	167	6	0.4
Birds	116	1029	9.5	1
Mammals	59	505	4.5	1
All plants	584	22137	240	0.2
Gymnosperms	2	242	0.8	0.3
Dicotyledons	120	17474	190	0.06
Monocotyledons	462	4421	52	0.9

After Smith et al. (1993).

so the approximation seems reasonable. Extinctions have been recorded over the last 400 years with some degree of reliability. With the background extinction rate of 5/year, we should expect about 2,000 extinctions in 400 years. Assuming the total number of species is 10 million, and that the extinction rates are no higher than the background rate, we can calculate the expected number of extinctions for various taxonomic groups (Table 7.2). For example, among 9,500 birds, we would expect about 2 extinctions in 400 years (9,500/10,000,000·2000 = 1.9), whereas the observed number of extinctions is 116. The table was limited to three taxa, mammals, birds and molluscs, for which there is good data.

Obviously, looking at Table 7.2, we have observed many more extinctions than expected, if the simple assumptions embodied in the calculations are correct. The calculations for the expected number assume that the current extinction rate is approximately the same as the background rate of extinction in geological time (between 1 and 5 species globally per year), that extinctions occur with more or less equal frequency among different taxa, and that there are 10 million extant species. It is worth noting that many uncertainties in the number of observed extinctions suggest that these values

Table 7.2. Observed number of extinctions globally in the last 400 years (from Table 7.1) and predicted number of extinctions in the same period, assuming a total of 10 million species and that the current rates are equal to background rates in geological time.

Taxon	Number of species (x1000)	Expected number of extinctions if background rate = 1/year	Expected number of extinctions if background rate = 5/year	Observed number of extinctions
Mammals	4.5	< 1	1	59
Birds	9.5	< 1	2	116
Molluscs	100	4	20	191

are likely to be underestimates; there is no doubt that we will not have noticed the loss of at least some taxa, even among the best studied groups. The result would be an even greater disparity between the observed and predicted number of extinctions, were they to be corrected.

There are two explanations for the disparity. Either extinctions in the last 400 years have occurred almost exclusively in the three taxa for which we have good data, namely mammals, birds, and molluscs. Or, the current extinction rate is at least two orders of magnitude higher than background. Other evidence for elevated extinction rates was examined by Smith and colleagues in 1993. They found that any taxonomic group or geographic region that is poorly studied will appear to be in good health because extinctions are difficult to observe and known extinctions will be few. For example, almost all the vascular plant extinctions recorded in Africa have been recorded in South Africa. However, it is difficult to know just how much of the difference between countries is because of different taxonomic and monitoring effort, and how much is due to the impact of developed economies on their environment. Records from the Indonesian island of Sulawesi indicated that the Caerulean butterfly (*Eutrichomyias rowleyi*) had not been seen for several decades, there were no recent records for many of the endemic species of the fish family Adrianichtyidae, and at least seven endemic bird species had not been observed for more than a decade. Yet none of these species were recorded in international threatened species lists. Only one bird species is listed in international data bases as extinct on the Solomon Islands, but Diamond and colleagues reported that 12 species have no definite records since 1953 and islanders report that several of these have been eliminated by cats. Similarly, there are no recorded extinctions of fishes on the Malay Peninsula, but only 122 out of a total of 266 previously described species of freshwater fish were found over a four-year period.

Even if we make the very conservative assumption there have been no unobserved extinctions of mammals, birds, or molluscs in the last 400 years, current rates in these three groups are still substantially above background. We are likely to be in the midst of an mass extinction event of a magnitude matched only by five other such events in geological time.

Given the kinds of uncertainties described above, it is unlikely that we can estimate the rate of species extinctions accurately. Different attempts to estimate the rate have resulted in quite similar values, but it is not possible to know how independent they are of one another. Only a very small proportion of recorded extinction events since 1600 are from continental tropical forest ecosystems. Most recent projections for future species loss take into account expected loss of tropical forest. Such estimates are uncertain mostly because the necessary information on numbers of species, population sizes, distribution, and the kinds of impacts are themselves uncertain.

7.2.3 The Causes of Extinction

The loss of the last few individuals of a species, while regrettable, is of less interest than establishing the causes that lead the species to become so reduced in the first place. In very small populations, it is likely that demographic and genetic processes play a major role in determining the fate of a species, together with environmental variation and other stochastic and deterministic factors. It is clear in many instances that the direct, ultimate cause of population decline has been the activities of humans. These activities include:

- habitat destruction and fragmentation
- overexploitation (overharvesting)
- pollution
- introduction of exotic species (that frequently become competitors or predators of native species)
- global climate change

The ecology of certain species makes them more vulnerable to extinction (Pimm et al. 1988; Pimm 1991). Species that are locally rare, geographically restricted, or limited to a narrow niche may be prone to decline. Species that are variable within their range, or are variable in time, may likewise be susceptible to global, permanent reductions in population size. Short-lived species (with a short generation time) may have a higher risk of extinction per year, although they may have similar per-generation extinction risks as longer-lived species. Species with slower growth rates may take longer to recover from population reductions.

In addition to these general characteristics, there are several properties that may make a species especially susceptible to human impacts.

Habitat overlap: Species may be threatened because they are tied to the same types of habitat that are preferred by people. The biota of relatively accessible areas with fertile soils and benign climates are subjected to agricultural and urban development. Similarly, the biota of coastlines, major rivers, and streams are subject to waste disposal, urban development, and the impacts of transportation. The consequences of human use of freshwater systems are especially important in arid environments such as Australia.

Harvesting: Species that are palatable or otherwise valuable to humans are susceptible. The most spectacular examples of animal species' declines and extinctions usually involve at least some harvesting pressure. Whenever species are harvested at levels above the maximum sustained yield, the species will be driven to extinction by systematic (deterministic) pressure, quite apart from increased risks that result from chance events. Species inhabiting small oceanic islands have been relatively susceptible to hunting pressure.

Home range requirements: Animals with extensive home ranges frequently occur at low densities, and they are likely to be susceptible to the changes in human dominated landscapes, including reduction in the area of available habitat and fragmentation of the remainder into a large number of relatively small patches (see Chapter 6).

Limited adaptability and resilience: If a species has limited dispersal capabilities, limited reproductive capacity, or narrow and inflexible habitat requirements, then it is unlikely to be capable of rapid recovery from disturbance and is likely to be relatively prone to extinction in human modified landscapes.

None of these characteristics are perfect predictors of a species' vulnerability to extinction. Species with similar ecological characteristics may survive for very different lengths of time. This is because extinction risks depend on all these general characteristics, as well as several others that are specific to particular populations and landscapes. We analyzed many of these in previous chapters: density dependence, age and sex distribution of individuals within a population, correlation of environmental variation, dispersal barriers, and habitat corridors are some of the factors that interact with species-specific factors. Population viability analysis can be described as the use of models to combine all relevant factors in the evaluation of extinction risks.

Given that you have built a reasonable model and estimated extinction risks of a species, how do you interpret the results? If your model predicted a 10% risk of extinction in 100 years, is this bad, or is it acceptable? Some scientists believe that the interpretation of results at this level is a political process that requires criteria imposed by society rather than by the scientific community alone. Others believe that scientists nevertheless have a responsibility to provide guidance. International conservation organizations have been using a set of categories for ranking the conservation status of threatened species, which we will discuss next.

7.2.4 Classification of Threat

When biologists communicate about the risks faced by different species, they use a set of qualitative categories for different degrees of threat. These categories go under the heading of the conservation status of a species. Terminology for conservation status has been developed principally by agencies for conservation and environmental impact assessment such as the International Union for the Conservation of Nature (IUCN), based in Gland, Switzerland. Classifications of threat are important because they form part of the basis for the determination of priorities in threatened species management. They provide an index of species that are most likely to become extinct in the near future.

The IUCN in 1994 defined the following set of categories for the conservation status of taxa at the level of species and below.

Extinct: There is no reasonable doubt that the last individual has died.

Extinct in the wild: Known to survive in cultivation, in captivity, or as a naturalized population well outside its past range. Exhaustive surveys have failed to record an individual.

Critically endangered: Facing an extremely high risk of extinction in the wild in the immediate future.

Endangered: Facing a very high risk of extinction in the wild in the near future, but not critically endangered.

Vulnerable: Facing a high risk of extinction in the wild in the medium-term future, but not endangered.

Lower risk: The taxon has been evaluated and does not qualify for any categories above.

Data deficient: There is inadequate information to make a direct, or indirect, assessment of its risk of extinction based on its distribution or population status.

Not evaluated: The taxon has not yet been assessed against the criteria.

These categories are supported by decision rules related to range, population size, population history, and extinction risk. Mace and Lande (1991) were the first to propose quantitative, risk-based criteria for these categories. These risk-based criteria defined the three threatened species categories as follows:

CRITICAL: 50% probability of extinction within 5 years or 2 generations, whichever is longer

ENDANGERED: 20% probability of extinction within 20 years or 10 generations, whichever is longer

VULNERABLE: 10% probability of extinction within 100 years

These criteria are in terms of time and risk of extinction, so a time-to-extinction plot (Figure 2.8 in Chapter 2) can be used to represent these risk-based criteria. On such a plot, each category will be represented by a single point.

For example, Figure 7.3 (based on Akçakaya 1992) shows three time-to-extinction curves. The category definitions are represented by a triangle (for CRITICAL), a square (for ENDANGERED), and a circle (for VULNERABLE), for a species with a generation time of less than or equal to two years. The curves were produced by three models that differed only in terms of the standard deviation of the growth rates, which were specified so that each curve goes through one of the points representing the criteria.

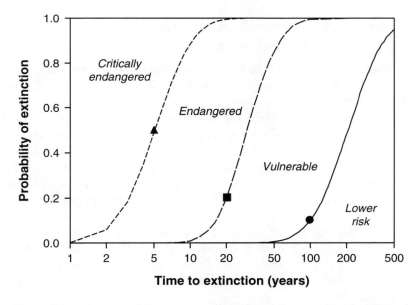

Figure 7.3. Levels of threat as a function of time and probability of extinction (based on Akçakaya 1992; note that the horizontal (time) axis is in a logarithmic scale).

Even with a set of universally accepted categories of threat, assigning species to categories based on their biology and demography remains a difficult task, especially for species about which we have little information. In addition to the above risk-based criteria for threat categories, Mace and Lande (1991) proposed a set of criteria based on population size, subdivision, variability, and recent declines. Which type of criteria should be used for assigning a particular species to threat categories depends on which type of criteria is more reliable, which in turn depends on what is known about the species. When quantitative information about a species is available, then population viability analysis provides a reliable tool for using the risk-based criteria.

7.3 Components of population viability analysis

There is no single recipe to follow when doing a PVA, because each case is different in so many respects. In this section, we will discuss some of the main components that a PVA might have. Not all PVAs will have all these components, and some will have others that are not discussed here.

7.3.1 Identification of the Question and Estimation of Parameters

Any scientific inquiry starts with a question, and population viability analysis is no exception. Although the question or problem to be addressed might seem to be obvious, it is nevertheless important to state it explicitly. This is because the question is likely to change in the course of a PVA. Initially, the question might be very general, such as "Is this species threatened, and if so, why?" The less we know about the species, the more general the questions will be. At this step (Step 1 in Figure 7.4) a PVA should concentrate on the identification of factors (including natural factors and human impacts) that are important in dynamics of the specific populations under study, as well as conservation and management options. The methods to be used for this depend on the specific case at hand, and might include statistical analysis of historical data, comparison of populations that are declining with those that are stable, and correlating recent changes in the environment (climatic or habitat changes, introduced species, changing harvest patterns, etc.) with changes in the species.

After the available information about the ecology of the species and its recent history is collated and reviewed, the questions are likely to become more specific. Examples of such questions include:

What is the chance of recovery of the Spotted Owl from its current threatened status?

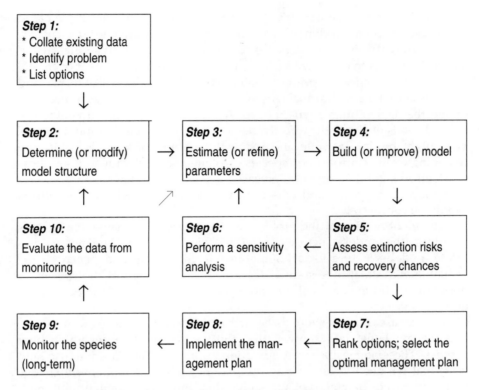

Figure 7.4. Components of a population viability analysis.

What is the risk of extinction of the Florida Panther in the next 50 years?

Is it better to prohibit hunting or to provide more habitat for elephants?

Is captive breeding and reintroduction to natural habitat patches a viable strategy for conserving Black-footed Ferrets? If so, is it better to reintroduce 100 Black-footed Ferrets to one habitat patch or 50 each to two habitat patches?

Would translocation of Helmeted Honeyeaters from their current populations to empty habitat patches minimize the extinction risk?

Is it better to preserve one large fragment of old-growth forest, or several smaller fragments of the same total area?

Is adding a habitat patch to the reserve system better than enhancing habitat corridors to increase dispersal among existing patches?

We will discuss the components of PVA with a hypothetical example based on the metapopulation dynamics of the California Gnatcatcher (*Polioptila c. californica*), which is a threatened bird species. The California Gnatcatcher has declined due to extensive agricultural and urban development of coastal sage scrub, the species' primary habitat type in southern California and northwestern Baja California (Atwood 1993). It was listed as threatened under the U.S. Endangered Species Act in 1993. For our hypothetical example, let's evaluate the effectiveness of a specific conservation measure. This might be, for example, increasing the amount of suitable habitat by removing exotic (introduced) species of plants. So, we want to know how much effort should be spent to increase the persistence chance of the species over the next 50 years, and where (in which habitat patches) these efforts should be concentrated.

The identification of the problem and the specific management options determine the model structure to use (Step 2 in Figure 7.4). The most appropriate model structure for a population viability analysis depends on the availability of data, the essential features of the ecology of the species or population, and the kinds of questions that the managers of the population need to answer.

In our case, the question concerned which habitat patches to improve, and the available data showed that the species lives in several habitat patches. These suggest a metapopulation approach. The question also suggests that we need to know how the parameters of the model will change with improved habitat. Let's assume that improved habitat will both increase the growth rate of the population, and its carrying capacity.

The next step is to estimate the model parameters with field studies (and sometimes experiments). The kind of parameters that need to be estimated will depend on the model structure, and the type of data already available. In our example, we need to first determine the geographic configuration of habitat patches (Figure 7.5) and their carrying capacities. We also need to know the growth rate and its variation, the correlation of environmental fluctuations and the rate of dispersal among populations. In addition, we need to find a way to relate the amount of management effort (e.g., the area from which the introduced plants are removed) to improvements in the population parameters (growth rate and carrying capacity).

For most PVA studies, this is the limiting step, because data are often insufficient. However, if a decision will be made no matter what, it is better if the decision-maker has some input from a PVA, even if the data are not perfect. If a parameter is not known very well, then a range of numbers can be used for that parameter instead of a single number. For example, if the average dispersal distance of California Gnatcatchers is about 3 km, but is

Components of population viability analysis 227

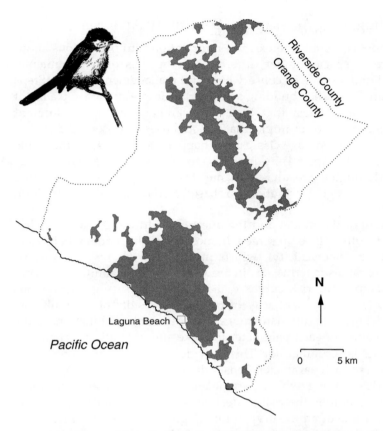

Figure 7.5. The spatial structure of a California Gnatcatcher (*Polioptila c. californica*) metapopulation in Orange County, based on the distribution of suitable habitat (after Akçakaya and Atwood 1997).

not known accurately, we can use a range of 2 km to 4 km. These ranges can be used in a sensitivity analysis (see below), similar to the one you did in Exercise 2.4 (in Chapter 2).

When there is not enough data for a particular rare or threatened species, some studies use data from a more common (thus better studied) species in the same genus or family. In some cases this may be reasonable, but only if the "borrowed" data are limited to general life history characteristics, such as whether to use age or stage structure. It does not make sense if the data include vital rates or numerical values of any other parameter. If the two species were so similar that you could use vital rates from one species to model the other, it is unlikely that one would be rare or threatened and the other one common.

7.3.2 Modeling, Risk Assessment, Sensitivity Analysis

Building a model combines the existing information into predictions about the persistence of species under different assumptions of environmental conditions and under different conservation and management options (Steps 4 and 5 in Figure 7.4). When building a model, it is important to keep a list of assumptions made. Models that look very similar may have different assumptions. Suppose, for example, that a model uses an observed distribution of individuals among age classes that happens to be close to the stable age distribution (Chapter 4). Another model *assumes* a stable age distribution. Although the result is the same, it is still important to remember that the first model's age distribution was based on data, and the second's on an assumption.

The structure of the model and the questions addressed usually determine how the results will be presented. In most cases, the model will include random variation (stochasticity), which means that the results must be presented in probabilistic terms, i.e., in terms of risks, probabilities, or likelihoods. For example, the risk curves that we have been using in previous chapters provide a convenient way of presenting results of a simulation. Often, the model must be run many times, with different combinations of the low and high values of each parameter to make sure that all uncertainty in parameter values is accounted for. This provides a way to measure the sensitivity of results to each parameter. Sensitivity analysis (Step 6 in Figure 7.4) is useful for determining which parameters need to be estimated more carefully. If, for example, the risk of decline is very different with the low value and high value of adult survival rate (Figure 7.6), then the results are sensitive to this parameter, and we can conclude that future field studies should concentrate on adult survival rate in order to estimate it more accurately. This feedback from modeling to field work is represented by an arrow from Step 6 to Step 3 in Figure 7.4.

7.3.3 Cost-benefit Analysis

When simulations include those with different management options, sensitivity analysis also gives information about the effectiveness of these options. For example, if we identified 3 patches where habitat for the California Gnatcatcher could be improved by human intervention, we could simulate the effect of improving habitat at each of these patches, plus at each pair of them, plus at all three. This would give us 7 alternative management actions, in addition to the option of no action. We could then rank them in order of increasing effectiveness. For this example, we might expect that the larger the area where habitat is improved, the lower the extinction risk of the gnatcatchers. The obvious choice is to improve the habitat in all three patches. In

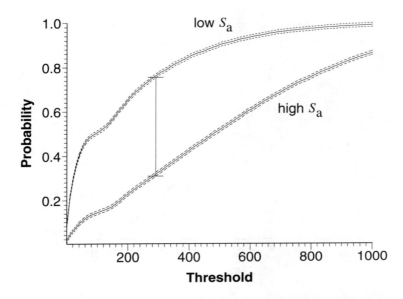

Figure 7.6. Sensitivity of the risk of decline of the California Gnatcatcher metapopulation to adult survival (S_a). The curves show the probability of falling below the threshold within the next 20 years with low and high estimates of S_a. The vertical line shows the largest difference between the two risk curves (based on Akçakaya and Atwood 1997).

reality the choices are much less obvious, because improving all three patches may cost more than what is available for California Gnatcatcher habitat management; this means we need to consider the costs as well. We could rank the 8 options with respect to both their benefit (reduction in risk of extinction) and with respect to their cost (Figure 7.7).

It is important to note that, in this graph, the cost of each management option (in units of currency) and its benefit (in units of risk) are in different axes. If, instead of analyzing the effect of habitat management on a threatened species, we were analyzing the expected return from different investments, both cost (investment) and benefit (expected return) could be expressed in monetary units (after perhaps some modification to account for uncertainties, inflation, etc.). We could then divide the benefit by the cost for each investment option and find the option with the highest benefit:cost ratio. This approach might also be applicable in the management of natural resources where the extinction of the resource is not likely (we will discuss such cases further in the next chapter). We cannot do this in the present case, because it is not possible to put a monetary value on the existence (or extinc-

230   Chapter 7 Population Viability Analysis

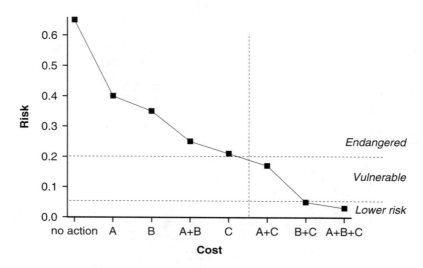

Figure 7.7. Risk of extinction of a hypothetical metapopulation with no management ("no action" option), and under 7 management options (which involve improving the habitat in one, two, or all three of the three habitat patches, A, B, and C). The options are in order of increasing cost from left to right.

tion) of a species, much as it is not possible to put a monetary value on human life. Although there are attempts to do just that, it is obviously a matter of value judgment, and each person will make the judgment in a different way. The way we used the benefit and cost information in Figure 7.7 (by keeping the units separate) sidesteps this problem of subjectivity by deferring it (as we shall discuss below).

A cost-benefit plot such as Figure 7.7 can be used in two different ways. First, the maximum amount of money that could be spent might be fixed (represented by the vertical dotted line). In this case the analysis could be used to select the option that gives the lowest risk. Conversely, the risk level that is targeted might be fixed (represented by the horizontal lines that correspond to IUCN's risk-based criteria). In this case, the analysis could be used to select the least costly option that meets the target. The target might be, for example, to move the species from the "endangered" category to "lower risk" category. In either case, the results of modeling are used to select the optimal management plan (Step 7 in Figure 7.4), under the particular set of conditions.

Different types of management might be represented by different curves. For example, we might also have the option of decreasing the risk faced by the gnatcatchers by reducing the cowbird populations that parasitize their nests. As we discussed in the previous chapter, cowbirds lay their eggs in the nests of other birds, which then raise cowbirds instead of their own young. If we had the option of removing cowbirds from a number of patches, we would end up with a different cost-benefit curve. These curves might intersect, which means that the best management option (habitat improvement, cowbird removal, or a combination) might differ depending on the total amount of resources available for management. We will explore this further in the exercises.

As you might have noticed, the selection of the management option carries subjective value judgments, even though we tried to avoid them. Somebody still must decide how much money to spend to reduce the risk faced by gnatcatchers (as opposed to other species, or, say, as opposed to improving water quality for human consumption). Or, somebody has to decide what the target risk level will be. However, these questions are clearly outside the scope of PVA, or any other scientific and technical analysis. Population viability analysis can be used to inform the decision-makers, politicians, and the public about the consequences of various actions and nonactions, but cannot (and should not) be used alone to make these societal decisions.

7.3.4 Implementation, Monitoring, Evaluation

With the selection of the best course of action under a given set of conditions (Step 7 Figure 7.4), the function of modeling is completed, but only temporarily. The next step is the implementation of the plan (Step 8). It is important that the field studies continue during and after the implementation to monitor the species (Step 9). The results of monitoring can give valuable information about the response of the species to management, as well as provide more data to refine model parameters and improve the model (Step 10). For example, we might discover, upon evaluation of demographic data after the implementation of the plan, that gnatcatcher vital rates increase faster than predicted in response to removal of cowbirds, but the carrying capacity responds slower than expected to the improvement of the habitat. Such a finding would definitely require modifying the model, refining its parameters, and re-estimating the extinction risks under different management options.

7.4 The limits of population viability analysis

Managing risks for natural populations is concerned with allocating resources to wildlife conservation. These resources are scarce because commitment of resources to conservation results in economic trade-offs to the rest of society. If our decisions are emotive, we face the possibility of inefficiency in using resources with the consequent loss of species that might otherwise have been avoided. The best way to address this problem is to apply methods that result in quantitative evaluation of risks for natural populations. We can use these results to underpin management decisions.

Thus we might, by consensus, define an acceptable level of risk for the extinction of species. We could then use this benchmark to help allocate conservation resources. An assessment of the risks faced by species will tell us if they are above or below the acceptable level of risk and if the risks are too high, we will manage the population in such a way as to reduce the risks.

As we discussed above, decisions involving "an acceptable level of risk" or "benchmark risk" are outside the scope of PVA. Population viability analysis can, for example, inform the public about the risk faced by the northern spotted owl if the logging of old-growth forests continues, versus the risk if the logging is stopped. But it cannot be used to compare the trade-off between the long-term persistence of a species and the loss of logging jobs, or to compare the responsibilities of the society to those who loose logging jobs versus those laid off by an airline, telephone or computer company. However, other methods (related to modeling harvested populations) we explored in previous chapters might be used to explore whether the logging jobs would be lost even with continued logging, albeit 5 or 10 years later. We will further explore decisions involving the management of natural resources in the next chapter.

Mark Shaffer, in a thesis published in 1981 that is one of the foundations of modern conservation biology, distinguished between "systematic pressure" and "stochastic perturbation" as causes of population extinction. Graham Caughley in 1994 differentiated between the "small-population paradigm" dealing with stochastic influences on small populations, and the "declining-population paradigm" dealing with the (largely deterministic) ecological causes and cures of population decline. He suggested that the principal contribution of the small-population paradigm is the theoretical underpinning it provides to conservation biology, even though the theory bears tenuous relevance to the problem of aiding a species in trouble.

This claim has been used to suggest that population viability analysis is limited to modeling stochastic influences on small populations, and ignores the ecological causes of most systematic declines by assuming constant or stationary conditions. Mark Boyce in 1992 noted that the distinction between deterministic and stochastic processes is artificial because all ecological processes are stochastic. Both processes can and should be included in models. The processes of systematic pressure, thought of as "deterministic," can be modeled in the stochastic models we have developed in previous chapters, with a growth rate that is less than 1.0, or a carrying capacity that is declining in time, or both.

Obviously models can and do incorporate deterministic decline, human caused factors, external influences and the effects of habitat loss. The challenge that population ecologists take when they build a model is to express all these effects in terms of the viability of the species, expressed for example as the risk of extinction. The set of ecological factors that can be included in a model is limited only by data and by one's imagination. If an ecologist has ideas concerning the forces that govern the chances of persistence of a population, then they can and should be included in the model, irrespective of their origin, and irrespective of their deterministic or stochastic nature (Akçakaya and Burgman 1995).

As Caughley (1994) correctly pointed out, no modeling effort by itself can determine why a population is declining or why it has declined in the past. For modeling to be successful in evaluating options for management of species, it must be part of a larger process of PVA, which must incorporate other methods, including study of natural history, field observations and experiments, analysis of historical and current data, and long-term monitoring. Any model entails assumptions about the ecology of a species. A model may assume that some or all of the mechanisms generating historical data remain unchanged in the future. If the model structure is incorrect or inappropriate for the species in question, serious errors in prediction are likely. Errors, together with uncertainties, are magnified into the future with each time step, so usually only a few time intervals can be predicted with any certainty. The omission of an important process such as loss of habitat, competition, or predation from introduced species, impacts of disease or parasites, or the impacts of rare catastrophic events, may substantially affect what is best to do to manage a population to avoid extinction. The ecology of species and the role of management should be, in the words of Mark Boyce (1992), the nuts and bolts of modeling exercises.

Of course, such considerations raise the issue of data availability. Data deficiencies plague attempts at building models aimed at solving real-world problems. Even the simplest models require more parameters than are usually available. On the other hand, frequently we need to incorporate a multitude of factors (related to, for example, the behavior of and future changes in human populations and their effects on habitat). We will never have a "complete" data set for any species. However, incomplete information does not mean that meaningful results are impossible to obtain. For there is very significant value in building a model for its own sake. It clarifies assumptions, integrates knowledge from all available sources, and forces us to be explicit and rigorous in our reasoning. It allows us to identify, through sensitivity analyses, which model structures and parameters matter, and which do not. It results in a set of logical statements that are internally consistent, and it allows us to explore the consequences of what we believe to be true, even in the absence of relevant, complete data. The only rule is that people who use a model (whether computer-implemented or not) should be aware of its assumptions and limitations, and communicate these together with the results.

The human population makes itself felt largely through the destruction of habitat of other species. The consequent decrease in natural population sizes adds to the other factors that tend to drive species to extinction such as competition, predation, disease, extreme environmental conditions, and the deleterious effects of inbreeding in small populations. Risk assessment and PVA are essential if we are to allocate scarce resources to conservation and wildlife management as efficiently as possible, thereby minimizing the number of species that will become extinct.

7.5 Exercises

Exercise 7.1: Habitat Management for Gnatcatchers

In this exercise, you will analyze the effectiveness of a management option for the California Gnatcatcher metapopulation we discussed in this chapter. The exercise is based on a model that is simplified from Akçakaya and Atwood (1997); it does not include several features of the original model, such as stage structure, catastrophes (fires and harsh winters), density-dependent dispersal, and Allee effects. The exercise is meant only as a demonstration of the concepts and methods explored in this chapter.

Step 1. Start RAMAS EcoLab, select the "Multiple populations" program, and open the file CalGnat.MP. Inspect the parameters for each population and the map of the metapopulation. Notice that most of the gnatcatchers are in two large patches, #5 and #10. Run the model. The simulation may run faster if you turn off the display of the map. If you have a very slow computer, you can stop at 300 or 500 replications.

Save the model and results (you can use the same filename). Investigate the results and record the risk of falling to 100 individuals within the next 50 years.

It might be difficult to read the precise value of the risk from the screen plot. Do the following to record this number precisely:

Click the "show numbers" button (the second button from left on top of the window), and scroll down the window to where you see "100" in the first column. Record the probability that corresponds to this threshold level. If "100" is not in this table, then click the third button on top of the window ("scale"). You will see a window with various plotting parameters (the exact numbers may be different in your simulation).

Title:	**Extinction/Decline**
☑ Autoscale *(checked)*	
X-Axis Label:	**Threshold**
Minimum:	0
Maximum:	762
Y-Axis Label:	**Probability**
Minimum:	0.00
Maximum:	1.00

First, uncheck the box next to "Autoscale" by clicking on it. (This makes the program use the values entered in this screen instead of automatically rescaling the axes.) Second, change the maximum value of the x-axis to the threshold (in this case, 100). Third, click OK.

Scroll down the table. The last line of the table will give the threshold (100), and the probability of reaching or exceeding that threshold. Record this probability (risk of falling to 100 individuals). This is the risk without any management.

Step 2. California Gnatcatcher has declined due to the loss of coastal sage scrub, its primary habitat type in southern California. As we discussed in the previous chapter, habitat loss usually results in discontinuities in the distribution of the remaining habitat. Often, one of the consequences of this fragmentation is increased edge effects. In this exercise, we will assume that increased edge effects cause increased parasitism of gnatcatcher nests by cowbirds (see Section 6.1.2), which causes a decrease in gnatcatcher population growth rate. The management option we will explore in this exercise is based on the possibility of increasing growth rate by the removal of cowbirds. How much increase in growth rate can be achieved depends on the amount of effort, which in turn determines the funds necessary for this management project. We will assume that the management effort is focused on the two largest patches, and the cost of the management program to achieve three levels of increase in growth rate in these populations is given by Table 7.3 below.

For each of the three options, estimate the risk of falling to 100 individuals within the next 50 years. To do this for each option,
(1) Start with the model you saved in the previous step.
(2) Select **Populations** from the Model menu.
(3) In the list on the left of the window, click on "Pop 5."
(4) Increase its growth rate (see table below).
(5) Select population 10, and increase its growth rate.
(6) Save the model under a different filename.
(7) Run the model and save the model again, this time with results.
(8) Estimate the risk of a decline to 100 individuals (see Step 1).
(9) Enter the risk in the table below.

Table 7.3. Cost of each management action and its effect on growth rates

Option	Growth rates of pop. 5 and 10	Cost (×$1000)	Risk
(no action)	0.98	0	
A	1.01	100	
B	1.04	200	
C	1.07	300	

Step 3. Based on your estimates in the previous step, which of these three options would you recommend:

(a) if the target of the management plan is to decrease the risk to less than or equal to 0.6 (i.e., 60%) with minimum cost?

(b) if the target of the management plan is to decrease the risk to less than or equal to 0.3?

(c) if the target of the management plan is to decrease the risk to less than or equal to 0.1?

Exercise 7.2: Comparing Management Options

In this exercise, we will use the gnatcatcher model we used above to compare the results of the previous exercise with those for another type of management. We will assume that this management plan divides the available resources between two types of management activities. First, it involves habitat improvement, which results in an increase in carrying capacities every year. We will model this using the "Temporal trend in K" parameter in the **Populations** screen (click the "Help" button for information on this parameter). Second, it involves cowbird removal as above, but only in years of high cowbird densities. The strategy is based on the assumption that the average growth rate is most effectively increased by preventing the lowest growth rates. Becuase it does not change all growth rates, this strategy cannot increase the average growth rate as much as the previous type of management, but it can also decrease the standard deviation of the growth rate. We will model this by increasing R and decreasing the standard deviation of R in **Populations**. All three types of changes are again restricted to the two largest populations. Table 7.4 gives the cost of this management plan at three levels of effort.

Table 7.4. Cost of each management action and its effect on dynamics of population 5 and 10.

Option	Changes in parameters of population 5 and 10			Cost (×$1000)	Risk
	Growth rate (R)	Standard deviation of R	Temporal trend in K		
D	1.00	0.30	2% of K	100	
E	1.01	0.28	3% of K	200	
F	1.02	0.26	4% of K	300	

Step 1. For each of the three options, estimate the risk of falling to 100 individuals within the next 50 years. To do this for each option,
(1) Start with the model in CalGnat.MP.
(2) Select **Populations**.
(3) Select (highlight) population 5.
(4) Increase its growth rate, decrease the standard deviation of R, and calculate and enter 2%, 3%, or 4% (see table above) of its carrying capacity for the parameter "Temporal trend in K" (e.g., for option D, calculate 2% of the carrying capacity, and type that number; don't type "0.02" or "2%").
(5) Select population 10, and make the corresponding changes (for "Temporal trend in K," use the same percentage, which should give a different number).
(6) Save the model under a different filename.
(7) Run the model, and save again, this time with results.
(8) Eestimate the risk of a decline to 100 individuals (see Step 1).
(9) Enter the risk in the table above.

Step 2. Use the following graph to plot the cost and the estimated risk for each of the 6 options (including the three options from the previous exercise). Connect the options A, B, and C with a solid line, and connect options D, E, and F with a dotted line.

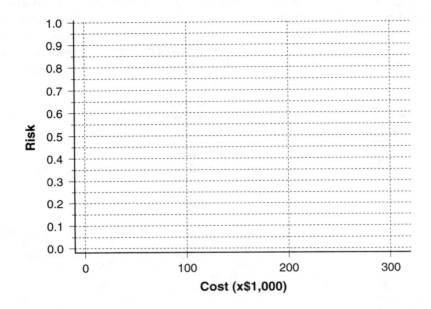

Step 3. Considering all 6 options, which would you recommend:

(a) if the target of the management plan is to decrease the risk to less than or equal to 0.6 (i.e., 60%) with minimum cost?

(b) if the target of the management plan is to decrease the risk to less than or equal to 0.3?

(c) if the target of the management plan is to decrease the risk to less than or equal to 0.1?

(d) if there is only $100,000 available for the management of gnatcatchers?

(e) if there is only $300,000 available for the management of gnatcatchers?

Exercise 7.3: Habitat Loss and Fragmentation

In this exercise, we will explore various ways the effects of habitat loss and fragmentation can be modeled in a PVA. The two-population metapopulation depicted in the left figure ("before") is fragmented by the construction of a road and becomes a three-population metapopulation (right figure, "after").

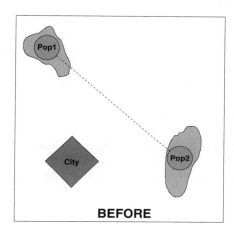
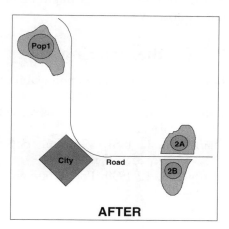

BEFORE   **AFTER**

Step 1. Start RAMAS EcoLab, select the "Multiple populations" program, and open the file 2-Pop.MP. Inspect the input parameters (under the Model menu) and the map of the metapopulation. Run a simulation, and record the risk of declining to 20 individuals (4% of total initial abundance). Save the file.

Step 2. Open the file 3-Pop.MP. Compare the input parameters and the map of the metapopulation to those of 2-Pop.MP. What are the differences? Run a simulation, and record the risk of declining to 20 individuals. Save the file.

Step 3. In the model in 3-Pop.MP, the sum of the initial abundances of the two new fragments is the same as before the fragmentation. The sum of their carrying capacities is also the same as before. However, this is not very realistic. The road construction does not just make two populations out of one population, it also reduces available habitat. Model the habitat loss as a 20% reduction in the carrying capacity and the initial abundance of the two fragments. Run a simulation, and record the risk of declining to 20 individuals. Save the model in a new file (i.e., use "Save As").

Step 4. Fragmentation can also cause a reduction in survival and fecundity of populations, because of edge effects. Model these as a reduction in population growth rate from 1.04 to 1.00 for the two fragments. Run a simulation, and record the risk of declining to 20 individuals. Save the model in a new file (i.e., use "Save As").

Step 5. What is the total increase in the risk of decline to 20 individuals, when all the factors are incorporated? How different is this result from a comparion based only on the number of populations? What other factors that we did not consider might be involved? How could these be incorporated into a PVA?

7.6 Further reading

Boyce, M. S. 1992. Population viability analysis. *Annual Review of Ecology and Systematics* 23:481–506.

Caughley, G. 1994. Directions in conservation biology. *Journal of Animal Ecology* 63:215–244.

Shaffer, M. L. 1981. Minimum population sizes for species conservation. *Bioscience* 31:131–134.

Shaffer, M. L. 1990. Population viability analysis. *Conservation Biology* 4: 39–40.

Chapter 8
Decision-making and Natural Resource Management

8.1 Introduction

This chapter, like the previous one, deals with aspects of applied population ecology that benefit from the application of population modeling and risk assessment. Our focus here is the role of uncertainty in environmental decision-making, and the management of natural resources. Population ecology is coming to terms with the fact that uncertainty and natural variability place a veil over all of our predictions. Some of this uncertainty is reducible and some is not, either because of practical constraints or because the source of the uncertainty is such that it is inherently unremovable. The community of people who make day-to-day decisions about wildlife and natural resource management are, by the nature of the environment in which they work, involved in risk assessment and risk management. In this chapter, we discuss how variability and uncertainty may affect decisions about environmental impacts and the management of natural resources, and the importance of dealing with economic and social factors in making natural resource management decisions.

8.2 Detecting impact

In this section, we consider issues related to the detection of environmental impacts. These issues have a slightly different focus than the modeling approach of the earlier chapters. Nevertheless, these issues have important implications for many problems addressed by applied ecologists. For example, if we want to model the viability of a species under a particular human activity, we need to know if the survival rates, fecundities, dispersal rates, or other model parameters are affected by this activity. This is done by analyzing data from impacted and nonimpacted areas and looking for differences. Such analyses often involve testing hypotheses. When an explanation for a phenomenon is proposed, a study is designed to test its veracity. For example, suppose that the cause of low egg count (a measure of fertility) in a fish population is thought to be pollution from a factory. A study designed to test this hypothesis might involve sampling fish from polluted and unpolluted rivers and counting their eggs. Suppose the mean egg count is lower in the polluted river. One possibility is that this difference is due to chance. These data are likely to show variation due to individual differences, so there will be some overlap between the measurements from the two rivers. The higher the variation (compared to difference between the means) the less reliable will be our conclusions. Thus, we might get two different means just by chance. We call this possibility the "null hypothesis." The other possibility is that pollution actually lowers fertility, and fish in polluted rivers have lower fertility. We call this explanation the "alternative hypothesis."

Because uncontrolled variation is present in the natural world, we apply an arbitrary criterion for such tests. If the chance of observing a result as extreme or more extreme than the results of the experiment is less than 5%, we accept that the results are "unlikely" to be the product of chance alone. In such cases, we are prepared to accept that the explanation (the alternative hypothesis) is true, and take a risk of up to 5% that the explanation is actually wrong. For example, the above study may conclude that there is less than 5% chance that the difference is due to chance, and not due to pollution. By considering this result to be "significant" in a statistical sense (i.e., accepting the hypothesis that fertility is lower in the polluted river), we are accepting a 5% risk of making the wrong conclusion.

Thus, one time in 20, we will accept the alternative hypothesis even though it is false, because one time in 20 the results we observe will be extreme by chance alone (this is called a Type I error). We could make the acceptance criterion more stringent, say, 1% or 0.1%. The reason that we are prepared to accept 5% is that we don't want to throw the baby out with the bathwater. If we set the acceptance level at 0.1%, we would reject many explanations that are in fact correct. Natural variation masks the effect of the explanation so that it is often difficult to substantiate something at small probabilities. For example, we may assume that fertility is not lower in the polluted river, even though it actually is. This kind of mistake is called a Type II error (Table 8.1). The "power" of an experiment to detect a given impact is inversely related to the probability of committing a Type II error; the more powerful the test, the less likely it is to erroneously conclude that there is no impact, when there actually is.

Table 8.1. Type I and II errors in an environmental impact assessment.

		State of the world	
		Impact	No impact
Result	*Significant impact detected*	correct	Type I error
of test	*No significant impact*	Type II error	correct

Decisions about the management, regulation, and conservation of populations frequently are based on statistical tests. For example, we test the hypothesis that the effluent from a factory has a detrimental impact on a fish population, or that timber harvesting adversely effects the survival rates in a bird population. If we accept the explanation (that there is an impact associated with the industry), it may result in the closure of the factory and a loss of jobs, or a change to more expensive harvesting practices with consequent costs for individual operators. If the conclusion was wrong (a Type I error), the curtailment of these productive activities may have been unwarranted. The local economic and social hardship that follows such decisions may have been avoided. On the other hand, we may reject the explanation and conclude that there is no impact. Life would go on and there may even be tacit approval for further development in the form of factory enlargement or more extensive harvesting. If this conclusion was wrong (a Type II error), there will be environmental degradation before the impact is detected.

In questions of environmental management, failure to reject the null hypothesis (no impact) is sometimes treated as synonymous with the conclusion that there has been no impact. If we fail to detect an impact, it may be that we haven't performed the appropriate experiments or did not have sufficient sample sizes. The result will be a Type II error. Such errors may involve biologically important outcomes that lead to things such as the collapse of a fishery or a substantial reduction in the natural distribution of a species that depends on old growth forest. Any erroneous conclusion is cause for concern, but traditional approaches to statistical inference concern themselves almost exclusively with Type I error rates.

The problem with the traditional approach is that it leads to a kind of blinkered view of environmental management. If we fail to see a problem, we conclude that there is no problem. The burden of proof lies with regulators such as Environmental Protection Agencies or with others who have an interest in protecting the environment. As a result, the rate of development and environmental impact may largely be determined by the availability of resources that are devoted to the detection of environmental problems. If there are more resources, larger studies will be conducted and more impacts detected. If resources are low, smaller sample sizes and insufficient monitoring will lead to fewer detections. An industry that undertakes its own monitoring studies (and there are many that do) will experience relatively few impediments to development if it does not allocate sufficient resources to assure a sound experimental protocol.

If an impact is detected, it is also relevant to ask if the impact is important. On the one hand, a statistically significant effect may be ecologically unimportant. In the above example, if the decrease in fertility due to pollution is small and the dynamics of the population are density-dependent, it may have no important consequences for either expected population sizes, or the risks of decline or increase in the population. On the other hand, a seemingly small effect may have important consequences for long-term viability. For example, a small decrease in the survival rate of breeding birds in a harvested forest may substantially increase the risk of extinction.

8.2.1 Power, Importance, and Significance: An Example

Consider the following example in which a government agency is charged with ensuring sustainable use of a natural forest. Sustainability is not clearly defined, but there are a few things that the public view as essential components of sustainability. One of these is the maintenance of representative populations of fauna that depend on the forest. A question is raised by a conservation group concerning the impact of harvesting on a threatened arboreal mammal. The industry suggests that it will modify its harvesting practices to accommodate the species. It plans to retain structural forest elements in a pattern and at a scale that match what is known about the niche requirements of the species. The government agency chooses to perform an experiment in which five areas are harvested using the modified techniques, and five areas are left untouched. Each area includes the territory of a single pair and each pair produces an average of one offspring per year. There are no direct estimates of survival rates but based on their body size and ecology, the animals are unlikely to live longer than 20 years. The investigator returns the following year and counts the number of offspring produced by the 10 pairs of animals. There is no statistically significant difference between the average numbers of offspring per territory in the harvested and unharvested areas. The investigator concludes that the experiment demonstrates that the new harvesting techniques are compatible with the conservation of the species in managed, harvested forests.

Is there anything amiss? By choosing to express the outcome as a "demonstration" of compatibility, the agency has assumed the mantle of the conventional approach to statistical inference. No impact was detected, therefore, there is no impact. If your job was to decide whether to permit harvesting in areas that are suitable habitat for the species, could you conclude that the modified harvesting techniques pose no substantial threat to the species? We can return to models as simple as those developed in Chapters 1 and 2 to answer this question. If we take the available information, we can assume that the fecundity is 1 offspring per pair and the initial population size is 5. We will ignore survival as the study is phrased in terms of fecundities, although we could assume survival is about 0.95. Figure 8.1 is the probability distribution of the number of offspring from 5 territories, assuming no adult deaths and no environmental variation.

The model accounts only for demographic uncertainty in the fecundity rate and ignores all other forms of uncertainty that may contribute to experimental noise and experimental error. Nevertheless, there is a 4% random chance that five pairs will produce a total of none or one offspring. There is a 12% chance that they will produce a total of fewer than 3 offspring. These probabilities are in the absence of any impact. Assuming that the forest harvesting activities affect only the mean fecundity rate, the study is likely to report a significant difference between the controls (the areas not harvested) and the treatments (the harvested areas) only if the five pairs in the treated areas produce a total of fewer than 2 offspring (because the probability of this event is 4%, which is just below the traditional significance level of 5%). If this happened, it would be considered "unlikely" to be the result of chance

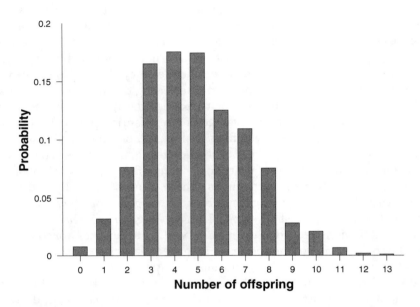

Figure 8.1. The probabilities of different total numbers of offspring from a population of five pairs of animals, assuming a mean of 1 offspring per pair sampled from a Poisson distribution.

alone, and we could, according to traditional protocol, accept the alternative hypothesis that the harvesting had affected fecundity. Otherwise, convention dictates that we should accept the null hypothesis.

The five pairs in the treated areas are likely to produce a total of none or one offspring, only if their mean fecundity is reduced a lot. In other words, the test is capable of detecting only the most extreme impacts. Even if the fecundity in the treated population were reduced by half, it is unlikely that the test would detect the impact. But we have seen in many of the models above that a 50% reduction in fecundity may have very important consequences for the expected population size and the chances of decline of a population.

Of course, if we had unlimited funds, we might increase the sample size in both the treatments and the controls, thereby improving the power of the test. Given a very large sample, the test may be adequate to detect very small changes in mean fecundity. The detection of a statistically significant impact would not imply a biologically important impact. A reduction in the mean fecundity of, say, 1%, may have no discernable impact on expected populations or chances of decline. Such an impact may be worth the rewards that flow from harvesting the timber, even in terms of nonutilitarian values. For example, royalties from timber harvesting may fund a program to eradicate a feral predator, more than outweighing the detrimental impacts of harvesting. Thus, a solution to the problem outlined above is complex. It involves consideration of the practicalities of increasing sample sizes, collecting further information, and weighing the costs of potential impacts on a threatened species against the costs of deferred access to a commercial resource. These have ecological, economic, social, and political ramifications.

Methods have been suggested that allow the joint consideration of Type I and Type II errors (see, for example, Bernstein and Zalinski 1983, Mapstone 1995). In doing so, we may specify the maximum impact that we are prepared to tolerate. This would, in turn, determine the experimental and monitoring effort necessary to determine if there is an impact. The details of such approaches are beyond the scope of this book. However, if such applications are to be made, it will be essential to develop ecological models that encapsulate the dynamics of populations so that the limits of acceptable impact may be established.

8.2.2 The Precautionary Principle

The above example emphasizes the need for a balanced approach to environmental management that accommodates the needs of society and the potential for damage to the environment. The notion of the cautious use of the environment is not a recent phenomenon. The precautionary principle was first defined and applied in West German environmental legislation in the late 1960s (WCMC 1992). Generally, it may be defined as "Where there are threats of serious or irreversible environmental damage, lack of full scientific certainty should not be used as a reason for postponing measures to prevent environmental degradation" (IGAE 1992). The principle implies a shift in the onus of proof from regulators and managers to developers proposing an action that may have an adverse impact on the environment, to demonstrate that the impact is absent, negligible, or worthwhile.

Some commentaries have suggested that obtaining proof that proposals would cause no damage is logically impossible, and that even the most ardent regulatory authority would be unlikely to make such a totally unattainable demand on developers. Almost all commentaries on sustainable environmental use agree that good management practices raise our chances of coping with unforeseen environmental threats and enhance our ability to maintain and improve the quality of existing environmental resources. The idea of the precautionary principle is not that proof be provided that there is no damage. Rather, the intention is to provide sufficient evidence that any impacts that do occur are likely to be within acceptable bounds.

8.3 Managing natural resources

In this section, we discuss various ways in which variability and uncertainty affect decisions about the management of natural resources—specifically, the management of harvested living resources. You have already dealt with harvested natural resources in previous chapters. For example, in Exercise 3.5, you explored the density-dependent effects of harvesting, and in Exercise 4.4, you simulated management strategies for a Brook Trout fishery. We begin our discussion with a hypothetical example that follows one developed by Adam Finkel in 1994.

Extractive reserves have been established in Brazil to promote sustainable development and resource conservation. These are natural areas where the government has granted rights to resident human populations to harvest such things as latex and Brazil nuts under established guidelines. Suppose you consider buying one hundred hectares of natural forest in Brazil for a total investment of $50,000. The area is a designated extractive reserve wherein you are obliged to harvest forest products sustainably (we will discuss this requirement later). You wish to hold onto the land for 50 years and you hope to make a profit. The hundred hectares supports a natural stand of the commercially valuable palm *Iriartea deltoidea*, and the stems can be sold for $10 each. There are 5,000 of these palms on your property. We'll assume that you will sell all the palms at the end of the 50 years. You also know that the environment is notoriously variable. While adult plants are more or less immune from variations in rainfall, the palms produce abundant seedlings in wet years. In dry years, the palms are much less likely to produce recruits. The question is, Is this a wise investment?

8.3.1 Predicting the Outcome

The agent for the sale argues that the natural variation in climatic conditions is beneficial for profit. The variation is so high that the growth rate of the population is 1.4 in good years and 0.7 in bad years. To simplify matters, let's assume there are only ever good or bad years, never any mediocre ones. Given this amount of variability, the upper 95th percentile for the number of stems at the end of 50 years is 168,162 (see Finkel 1994). The upper 95th percentile is a standard measure, telling you that the true value for the number of stems is unlikely to be higher than this, and is likely to be somewhat lower. At $10 per stem, there is a 1 in 20 chance that you could make more than one and a half million dollars. The investment provides you with the prospect of becoming relatively wealthy in your retirement. This is especially attractive when you think that this is not the absolute upper bound. There is a chance (albeit somewhat smaller) of making even more money.

It may be important to obtain a better estimate of the average expectation. You recall that the mean of a multiplicative process such as population growth is given by the geometric mean of the individual rates. Thus, taking the seller's scenario, a good year followed by a bad year will result in an overall annual growth rate, $R = \sqrt{1.4 \times 0.7} = 0.99$. On average, you will lose money on the investment. Each year, the number of palms will fall by an average of 1%. If this continues for 50 years, and there are 25 good years and 25 bad years, the overall rate of change will be $(1.4)^{25} \times (0.7)^{25} = 0.603$. This is the median (or, the most likely outcome) of the distribution of 50-year population growth rates. The number of palms at the end of 50 years would be $5000 \times 0.603 = 3,015$, with a total value of a little over \$30,000. Thus the value of the palms in 50 years time will be almost 40% less than it is today. It would seem to be madness to spend \$50,000 on something that, on average, loses \$20,000.

The agent does not give up, and offers the following argument: "It is true that half of the probability lies below \$30,000 and half lies above \$30,000. However, the point is misleading. The consequences if the number of stems exceeds \$30,000 far outweigh the consequences on the other side of the median. For example, it is equally likely that there will be either 30 bad years or 30 good years out of 50. But the windfall from 30 good years (number of palms = 96,182) far exceeds the loss if there are 30 bad years (number of palms = 94). One must weigh the chance of winning over \$960,000 against the chance of losing \$49,000. Both these numbers are equally likely. The correct way to look at the problem is to calculate the probability-weighted sum of the costs or benefits of the possible outcomes. After one year, there will be either 3,500 or 7,000 palms. The average of these numbers is 5,250; thus the population is expected to grow on average by 5% per year. How can you pass up an investment that on average will grow to \$573,000 by the time you plan to reap the rewards of the investment, sell up, and retire."

By this point, you begin to wonder what to make of all these arguments. Two of them make the investment seem a sure winner. The other makes it seem like a looser.

8.3.2 Explaining the Uncertainty

All that can be said is that the investment is volatile. One can be reasonably certain that the number of palms on the one hundred hectares will most likely be between 54 and 168,162 (5th and 95th percentiles; see Finkel 1994). Beyond that, it is not possible to be certain about anything. The mystifying and superficially contradictory nature of the various estimates is caused by the fact that the outcomes in the agent's model are lognormally distributed (Figure 8.2). When plotted on a logarithmic axis (as in this figure), the lognormal distribution has a symmetric shape; when plotted on a linear scale it would be skewed to the right. Because of the high variability, and the skewed shape of the lognormal distribution, point estimates (or single-number estimates) of the outcome, such as the median, the mean and the 95th percentile, may sound contradictory.

Once the kind of distribution of expected events has been characterized, it is possible to paint a complete picture of the potential risks and benefits of a decision. Thus, using Figure 8.1, we could create a table, or a curve, of the chances of there being less than a given number of palms. This way of looking at the problem can be achieved simply by creating a cumulative probability curve (Figure 8.3) by summing the values in Figure 8.2, in the same way that we created risk curves in Chapter 2.

With the two curves representing the probability distributions of population sizes at the end of 50 years, it is clear that all of the information given by the point estimates is accurate. Your estimate of the average outcome, 3,015 palm stems, is the median of the probability distribution, and it is also the geometric mean. The range of "reasonable certainty" (between 54 and 168,162 stems) is the region between the 5th and 95th percentiles. The arithmetic mean of the distribution is 57,300, a not unlikely event in the sense that it is within the region of reasonable certainty.

These curves provide additional information. The cumulative probability curve tells you that it is more likely than not that you will not recoup your $50,000 investment. In fact, the chance of making a loss is about 60% so the conclusion you reached after looking at the geometric mean was qualitatively correct. The agent's argument concerning the relative weights of different outcomes is also true. Any outcome in the top 25% of the distribution represents gains that at least outweigh the maximum loss of your $50,000 investment. The chance of doubling your investment is greater than 20%. The chance of a 10-fold increase is better than 5% and the chance of a 100-fold increase is about 1%.

There are no absolute rights or wrongs in making the decision. The wisdom of the choice depends on how you, personally, would react to the different possible outcomes. The question you should ask is, "How should I weigh a 60% chance of a loss against a 20% chance of a large gain?" If the $50,000 is all that is keeping you from the poor house and a life of misery, then the risk may not be worth it. If, on the other hand, the $50,000 is spare change, you may view the investment as you would a raffle ticket. The risk, in that case, might well be worth taking. You have to apply a kind of personal weighting factor. Costs and benefits, even when they both can be measured in terms of money, are not linearly related to the value of the investment. They are relative to your perception of, or the value you give to, potential losses and returns.

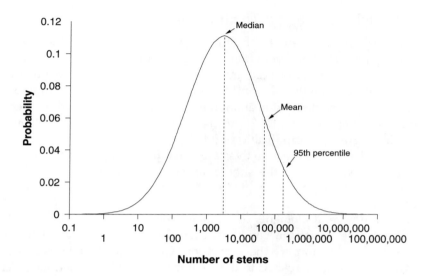

Figure 8.2. Probability density function for the number of palm stems expected at the end of 50 years, assuming equally likely years in which the growth rate of the population is either 0.7 or 1.4 (after Finkel 1994).

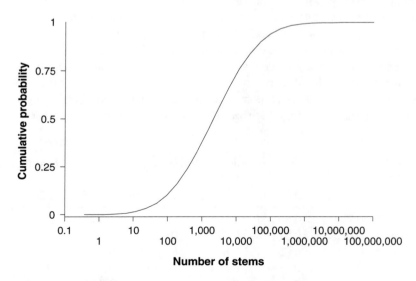

Figure 8.3. Cumulative probability function for the number of palm stems expected at the end of 50 years.

8.3.3 Model Uncertainty: The Importance of Detail

All of the arguments presented by the agent were true within the limits of the model in which there were only good and bad years. The problem was simplified so that it could accommodate available data (the average and variability of the growth rate), without making it unnecessarily complex. It served to inform the discussion about the effect of variability on the expected number of palms. In Chapter 2 we discussed the notion of model uncertainty, the degree of simplification and the specification of structures that relate the variable of interest (in this case, the number of palms) to the causal factors (the weather). Has the simple structure of the agent's model been misleading? Could a more detailed model improve your understanding of the problem?

There are several elements of biological detail that could be added to provide a degree of realism. You know that the assumption of good and bad years was made to simplify the discussion. Given that the production and survival of seedlings is related to the weather, it would be much more realistic to sample the variation in survival and fecundity from distributions that reflect the continuous nature of environmental variation. If the population were to decline, demographic uncertainty might play a role in the chances of various outcomes such as the loss of the entire population. Because the important component of environmental variation is concentrated on seedlings, the age or size structure of the population may be important. In 1993, Michelle Pinard published the results of her research into the population ecology of *Iriartea deltoidea*. Part of this research was a stage-structured model for the species (Table 8.2). The models you will develop in the exercises of this chapter, and the discussions in the following sections, are based on Pinard's work, but the scenarios for such things as market prices, levels, and patterns of natural variation are developed for illustrative purposes. This is not an accurate analysis for the species. It is intended only to provide examples of the ways in which population ecology interacts with decision making.

Table 8.2. Transition matrix for *I. deltoidea* from Altamira, Brazil (after Pinard 1993).

	<0.5 m	0.5–5 m	5–10 m	10–15 m	15–20 m	>20 m
<0.5 m	0.794	0	0	0.195	1.560	2.631
0.5–5 m	0.040	0.814	0	0	0	0
5–10 m	0	0.037	0.896	0	0	0
10–15 m	0	0	0.034	0.950	0	0
15–20 m	0	0	0	0.050	0.940	0
>20 m	0	0	0	0	0.045	0.828

The average growth rate of a population with these parameters is about 0.99, the same as the average growth rate in the example proposed by the agent. If we add the extra detail represented in Table 8.2, it allows us to explore some of the consequences of the assumptions that were made in the preceding discussion. We know from the simple model that variation may change your perception of the investment, particularly if you are not too averse to the risk of losing your money, and if the chances of a large gain are reasonable. One may add environmental variation to Pinard's (1993) model by specifying standard deviations of variation for the fecundity and survival rates. If the model is implemented on a computer, it is relatively easy to generate probability curves for the chances of population increase and population decrease during the next 50 years.

A modicum of biological intuition tells you that it probably won't be possible to fit 168,162 palms into 100 hectares. Palms that grow in densities greater than about 500 per hectare are very likely to compete for soil nutrients and light. This competition will be reflected in increased mortality, a consequence of self-thinning. Thus, not all of the possible outcomes suggested by the agent's model (Figure 8.1) are biologically plausible. Furthermore, this is a natural stand, and the palms are not spread evenly over the area as they would be in a plantation. They occur in clumps where microsite conditions provide the appropriate niche. Lastly, you know that trees are likely to self-thin, which can be modeled by a ceiling to the population. In the exercises at the end of the chapter, we will build a model that includes some of these biological details.

8.3.4 Strategies and Contingencies

Adding the biological details will make the model as plausible as it can be made, given the current state of your knowledge. This model can then be used instead of the simple (good-or-bad-year) model in evaluating the investment. When making this evaluation, you can explore three types of strategies for better managing this resource. These strategies may be used to improve the prospects for the investment, rather than accept the difficult choices that result from the scenarios above.

The first strategy concerns *decision rules* about the timing of the harvest. In the above discussion, we assumed that you'd harvest the palms at the end of 50 years. You could, instead, keep track of the number of palms each year, and decide to sell the population before the 50 years are up. If the population reaches a high value in 15 years, or 30 years, you could sell and take the profit early. For this, you specify an upper bound and say that if the population reaches this size, you will sell. In doing so, you state that there is a return on the investment that will satisfy you. You decline the possibility of even higher profit in return for reducing the chances that the investment will fail.

If you set the upper limit at, say, 20,000 plants, you are saying that a fourfold return on your investment is sufficient. The effect of the upper limit will be that you have no chance of making more than $200,000. The benefit of foregoing the opportunity of large profits is that the left hand side of the distribution, the chance of failure, will be much reduced. Because you are prepared to take profits early, there are fewer chances that you will fail.

The second type of strategy concerns the amount and distribution of the harvest. There are many possible harvesting strategies that may be employed to improve your chances of success. You could, for example, sell all of the palms immediately. This way you guarantee that you will break even, but you forgo any opportunity of making a profit. Another alternative is that you might harvest a small number of individuals each year, to contribute to running costs and to ensure that you take at least some return from your investment before the 50 years are up.

If you harvest each year, you also need to decide whether to harvest a fixed amount or a proportion every year. We know from earlier chapters that removal of fixed numbers of individuals generally destabilizes populations in variable environments, elevating risks of decline out of all proportion with the number of individuals removed. It is generally a better strategy to remove a fixed proportion of the population.

The third type of strategy concerns the management of the resource to improve its productivity. You may decide to invest additional capital to modify the processes to reduce the risk of failure, or to increase the expected size of the population. For example, you may invest in seedling stock, artificially increasing the size of the population at the outset, or in an irrigation system to water the plants whenever the weather is dry. We will explore these in the exercises.

The unconditional investment that was explored using the agent's simple model does not allow for making decisions (such as to harvest all palms before the 50 years are up) in response to the run of chance events. Nor does it allow the selection of alternative strategies related to harvesting regimes and additional investments to improve the productivity. In general, the presence of alternative strategies for management and responses to contingencies should make the decision about whether to buy or not somewhat easier. By judicious selection of rules and response mechanisms, it should be possible to develop a strategy that appeals to almost any investor.

8.4 The economic and ecological contexts of natural resource management

So far we dealt with uncertainties arising from the natural variability and lack of data. There are also other sources of uncertainty that originate from the social and economic conditions under which natural resource management decisions are made. In this section, we will discuss some of these factors. For instance, price uncertainty is important. The calculations above all assume that the palms will maintain the current net worth into the foreseeable future. Markets for commodities such as timber and ornamental plants are almost as volatile as tropical weather. The chance of increases or decreases in the relative price of the trees will weigh on any decision. Similarly, when making medium or long term predictions, there are other implicit assumptions such as the gamble that there will be no political upheavals that preclude you from accessing the assets when they mature.

The model for the palm trees deals with uncertainty in terms of the number of stems and ultimately, dollars. Even in such a straightforward scenario, the value of a dollar is not a simple thing. When making resource management decisions based purely on economic considerations, current worth is always considered against the discount rate, the factor by which future earnings are discounted to estimate their present value. The result may be a decision to take the entire population now (assuming reinvestment elsewhere for a higher rate of return), instead of harvesting on the basis of sustainable yield. This is probably the reason why open-access systems, such as the whaling industry, do not operate on a sustainable basis. Rosenberg and colleagues (1993) suggested that the solution to this problem is to recognize that property rights must be well defined and that rights imply duties and responsibilities. Several countries including Australia, New Zealand, Canada, Iceland, and the United States have recognized this problem and have granted individual quotas in some fisheries.

The value of an investment also depends on your attitude to the chance of losing your investment versus the chance of making a lot more dollars. People manage natural populations for many reasons other than their net market value. When values other than the purely utilitarian come into play, the role of uncertainty is magnified by the necessity to equate profit with other motivations. When Pinard (1993) presented the results of her study of the palm, she did so from the perspective of developing procedures for sustainable land use practices. She made the point that the concept of intergenerational equity is usually part of definitions of sustainability but that the concept is intractable from a land manager's perspective. For managed populations such as many of those described in this book, sustainability is measured in practice in terms of productivity, resource population stability, and yield maintenance. Pinard measured sustainability through population stability and expected yield continuity. The reserves are intended also to maintain genetic resources, forest structure, and associated ecological functions. If it turns out, for example, that harvesting latex, Brazil nuts, game species, and palms is not viable in the medium term of 50 to 100 years, the authorities would have to decide if it is worth continuing with the incentives and controls that make the extractive reserves possible. In making the decision, relevant factors would include such things as the relative value of ecological processes, genetic resources, conservation priorities, and alternative land use practices such as grazing. Because of the uncertainty in forecasting the economic value of alternatives and the difficulties inherent in relating very different values in the same currency, the decision would be determined largely by political and social forces.

8.4.1 Uncertainty and Sustainability

A collection of papers, published in the journal *Ecological Applications* in 1993 (vol 3, no 4, pp. 547–589) by Ludwig, Hilborn, Holling and others, discussed the sustainable management of resources when the future is uncertain. Uncertainty, for these authors, included environmental stochasticity, and economic and social change. Ludwig made the point that since flows from natural systems are limited, a conflict between human objectives and conservation of resources is inevitable, unless the rates at which humans use the environment are also limited. The objective of fisheries and forestry management in the past has been on maximum sustained yield, rather than on a yield that will ensure conservation of the resource. Fisheries managers have rarely been able to control the amount, distribution, and technique of fishing effort, even though such controls are necessary to achieve sustainable yield. There are many examples in which fisheries managers consistently allowed higher catch levels than indicated by the consensus of scientific advice. The temptation to increase yield at the cost of additional risk to the resource is often irresistible.

Ludwig and colleagues (1993) suggested that the exploitation of irregular or fluctuating resources is subject to the ratchet effect. During relatively stable periods, harvesting rates tend to stabilize at positions that are determined by bioeconomic systems that presume a steady state. A sequence of good years may encourage investment in infrastructure and capital. In sequences of poor years, the industry is likely to appeal to the government or the general population for help. Substantial investments and many jobs are at stake. Government response typically is direct or indirect subsidies. The ratchet effect is caused by the lack of inhibition on investment during good periods, and strong pressure not to disinvest during poor periods. The long-term outcome is a heavily subsidized industry that overharvests the biological resource on which it depends. There is no tradition of sustainable management in urban planning or development. The concept of sustainable management in agriculture is limited to stocking rates and water supply on individual farms, more closely related to maximum sustainable yield than to ecologically sustainable management.

A number of general principles were suggested in all of this discussion of sustainability and resource use. Human motivations and responses should be included as part of the system to be studied and managed, because human greed and short-sightedness underlie most difficulties in resource management. Claims of sustainability should be distrusted because past resource exploitation has seldom been sustainable, and frequently scientific advice is ignored. Resources should be managed explicitly for uncertainty by considering a variety of different strategies, favoring actions that are informative, reversible, and that are robust to uncertainty, and experimenting with the system and monitoring the results. Management strategies should be adaptive in the sense that uncertainty and surprise are an integral part of anticipated responses. Such an approach should be interdisciplinary and combine historical, comparative, and experimental approaches to resource use. Policies and actions are required that involve not only social objectives, but that continue to improve understanding and provide for flexibility in the event of surprises. Trial-and-error is often seen as an integral part of adaptive management. Use, monitoring, and the choice of reversible strategies will enhance our understanding of, and our ability to manage, natural systems.

All of the above discussions view management of renewable resources from the rather myopic context of single species or single, utilitarian values. Changing human values and social priorities form part of the context for resource management. Resource sustainability cannot be divorced from the sustainability of human economies, natural communities, and ecosystems. Sustainability is a moving target, not only because ecosystems change over time, but also because the economic, social, and political climates in which decisions are made change.

8.4.2 The Role of Applied Population Ecologists

Wherever there is uncertainty, there will be room for debate. Many decisions may appear intractable from a scientific point of view, but nevertheless they may be necessary from a pragmatic point of view. For example, when making decisions concerning the management of species, it is often necessary to involve expert judgements simply because no quantitative information is available on which to base decisions. Such judgments contribute to the priorities that are developed for allocating scarce conservation resources (IUCN 1995). When the decision affects competing demands on limited and environmental resources, the question immediately arises: "Whose experts?" Many natural resource management decisions require decisions based on forecasts that are inherently uncertain, and as we saw in Chapter 2, that uncertainty may take many forms.

There are no easy answers to the question "What is best?" It is the task of those involved in applied population ecology to present as complete a picture as possible to those who make decisions. The picture should incorporate mechanistic understanding, deterministic processes, stochastic variables, and the ensemble of uncertainties that contribute to the problem. A full treatment and careful presentation of the sources and consequences of uncertainty can make the search for an ideal solution easier. A mathematical model of a population is an explicit treatment of our understanding of the deterministic and stochastic mechanisms that affect that population.

The context in which the model is developed has profound consequences for its utility. Biological intuition is essential for constructing models but it is not sufficient. Decisions made by biologists without quantitative analysis are likely to lack rigor and consistency. Population models developed in isolation by a mathematician are likely to be biologically naive. Decisions based on realistic models, but made in isolation from bureaucrats, politicians, and interest groups are likely to be politically and socially naive.

The most important feature of applied population biology is that it be relevant to those that have to make decisions. Models are an important component in developing understanding and making predictions, and they are subject to the same caveat. Relevance may be determined simply by economic constraints. For example, if your task is to manage a plant population on a conservation reserve, the process of model development may be limited to a consideration of ecological dynamics and those management practices that can be brought into play within the limitations of a small budget. In other circumstances, the social context may be much more complex. A model of a population that has implications for the availability of significant societal resources is doomed to failure if it does not include stakeholders in its development. Irrespective of the abilities of a biologist or a modeler, the model will either lack the ability to answer the right questions, or those who must rely on its output will not have any confidence in it, or both. If the process of model building is collaborative and iterative, and if it involves representatives of all stakeholders, it has a chance of being useful.

Many models for natural populations are built in circumstances in which data and understanding are scarce. The relevance of models for environmental decision making is in the mind of the policy maker, and is not the realm of the modeler. An ecologist provides a service, a skill, and the end product is a set of recommendations that are bounded by assumptions and uncertainties. It is as important, if not more important, for the ecologist to communicate those uncertainties and assumptions, as it is to communicate the set of predictions. One of the reasons that models, and the people who build them, fall into disrepute is that models of complex or poorly understood systems will often produce different expectations. One of the reasons for this is that the bounds and the mechanics of the analysis are colored by what the model builder believes to be important. The fact that different models of the same natural system may generate different expectations is not surprising to modelers, but it is a source of frustration to decision makers. In such circumstances, the modelers may well have failed in their task because the creation of a sense of frustration implies that the sensitivities, limitations, and assumptions of the models have not been explained.

The use of models in decision-making should improve communication and understanding. If they do, the results they produce will be integrated quite naturally with value judgments and political constraints, to produce better decisions than could be made in the absence of models. To achieve these ends, models must be carefully and thoroughly documented, and limitations, sensitivities, and assumptions must be explicitly stated. Modelers must be sensitive to the needs and limitations of those people who intend to use them. Educational mechanisms that will allow modelers to develop the skills and experience necessary to produce useful models should be encouraged, as these are matters of professional responsibility.

8.5 Exercises

Exercise 8.1: Statistical Power and Environmental Detection

In this exercise, we assume you have taken an introductory course in statistics (or that your instructor has sufficient patience to teach you the fundamentals). We assume familiarity with hypothesis testing in general, and with t-tests and the calculation of standard deviations in particular.

Your role in this exercise is to monitor fish populations in a coastal management area and report on impacts of industrial activities. You know enough about the biology of a fish species to have developed a reliable model that incorporates density dependence. The carrying capacity, K, of the fish population is determined by the area of seagrass. There are two bays, one of which is a protected national park. You use this as a control. The other bay supports port facilities. You know from experience the growth rate, the survival rate, standard deviation of the growth rate, and the carrying capacity (K) in both bays in the absence of impacts.

There is a proposal to dredge part of the port to create a new dock, which will result in the elimination of about 10% of the seagrass in the bay, thereby reducing the carrying capacity for the fish population by 10%. You have enough money to conduct accurate censuses once each year. If you had five years of post-impact, good quality monitoring data, could you be reasonably confident of detecting the impact of the loss of seagrass, assuming that one in fact exists? We will use RAMAS EcoLab to help answer this question.

Step 1. Start the "Multiple Populations" program from the RAMAS EcoLab shell. Set the number of replications to 1, the duration to 5, and use demographic stochasticity. Set up two isolated populations. Call one the Park population and the other the Port population. Set the carrying capacity in the Park population to $K = 10,000$ and in the Port population to $K = 9,000$. For both populations, set initial abundance equal to K. Specify the following parameters for both populations:

Growth rate (R):	1.1
Survival rate:	0.5
Standard deviation of R:	0.11
Density dependence type:	**Scramble**

Make sure there is no dispersal or correlation between populations (However, in order to see both populations on the screen, you may want to set different coordinates for the two populations). Save your model.

Step 2. Run a single replication of this simulation. Select "Trajectory summary" from the Results menu, and write down the population sizes for the Park and Port populations from years 1, 2, 3, 4 and 5. To do this, first click the "text" button on the toolbar, then advance the population counter to 1 and then to 2 (population = 0 gives the total metapopulation abundance).

Step 3. For each population, calculate the mean (M) and the standard deviation (S) of the 5 population sizes. Use these four numbers to conduct a t-test of the differences between mean abundances at the two locations. You should do a one-tailed test because you expect the Port population to be smaller, on average, than the Park population. The formula for the test is

$$t_s = \frac{(M_{Park} - M_{Port})}{\sqrt{(S^2_{Park} + S^2_{Port})/n}}$$

where M_{Park} and M_{Port} are the average abundances of the Park and Port populations over the first five years of the simulation, S^2_{Park} and S^2_{Port} are the variances in the Park and Port populations, and n is the number of years ($n = 5$).

Step 4. Compare the value you calculate (t_s) against the t value of 1.86 (which assumes a Type I error rate of 5%). If the number resulting from your calculations is larger than this, standard protocols for hypothesis tests say that you may conclude that the difference between the average populations is unlikely to be due to chance. That is, you are free to conclude that there is a significant difference between the Park and Port populations. If the number you calculate is less than 1.86, you cannot reject the null hypothesis of no difference between the populations. That is, even though there is a true difference between the populations, you would have insufficient evidence to be sure.

In the simulations, you have specified a true difference between the populations. You may or may not have detected a significant difference between the two sets of numbers from the two locations, depending on the vagaries of the environment. If you were to repeat this exercise many times, you should find that, on average, you find a significant difference about 40% of the time and that you fail to detect a difference about 60% of the time. This failure rate is the type II error rate, the probability of failing to detect a true difference. Thus, even given 5 years of good quality data, a correct model, and a true decline of 10% in the carrying capacity, the monitoring program stands a better than even chance of failing to detect the impact. You might want to extend the length of time from 5 to 50 years, and then recalculate the test. Power should improve.

Models such as these may be useful in exploring alternative monitoring programs while exploring our assumptions about the population dynamics of a species. One way to do this is to calculate the power of a monitoring program by specifying different kinds of plausible impacts, different sampling designs, and by trying these experiments using a range of alternative possible models.

Exercise 8.2: Sustainable Catch Revisited

The object of one of the exercises in Chapter 4 was to estimate sustainable catch for a fishery in the presence of environmental variability. In this exercise, we will develop the issue a little further. Model structure is never entirely certain, and often there will be more than one plausible model for a population. The objective of this exercise is to develop a harvest strategy for a fishery population such that you achieve maximum harvest over a 20-year period without taking any important risks of loss of the population (there will always be some risk the population will be lost, even if there is no fishing or other impact).

Step 1. Start the "Age and stage structure" program of RAMAS EcoLab. Open the file called Cod1.st. This model represents a stage structured model for a cod population. There are 10,000 fish in the current population and the environment has a carrying capacity of 20,000. Only the stage 3 fish can reproduce (see the Stage Matrix under the Model menu). The population is regulated by scramble competition, and the maximum growth rate in the absence of density dependence is 1.3 (see Density Dependence under the Model menu). The standard deviation of each parameter is set to equal 10% of the mean (a modest amount of environmental variation).

The model is set up to run 1,000 replications of the model over 20 years. In this version, there is no harvesting. Run the simulation and examine the Trajectory Summary and the Extinction/Decline curves under the Results menu. The population tends to increase towards the carrying capacity of 20,000 and there is only a very small probability that the population will decline to fewer than about 9,000 individuals. We will use these results as the benchmarks against which to compare other model predictions.

Step 2. Open the file Cod2.st. It is a deterministic simulation for the same model, but with a harvest of 400 individuals per year taken from the oldest stage. If you run the simulation, you will see that this is the sustainable harvest from the population, in the absence of environmental variation.

Now load the file Cod3.st. This is the same model as in Cod2, except that the environmental variation present in Cod1 has been added back into the model. The total harvest is about 12,500 kg (see Harvest Summary, display text results, and scroll to the very end of the table). There are two important qualitative features to the results of this simulation. The first is that the average population declines (see Trajectory Summary). The second is that there is about a 10% probability that the population will become extinct. Quite apart from the ecological consequences of such an event, this represents a significant economic risk.

Step 3. Repeat this simulation, using fixed harvest amounts of between 200 and 600 individuals per year. This can be done by opening the Management and Migration sheet under the Model menu. Plot extinction risk versus number of individuals harvested. Record the total harvest from each simulation.

Step 4. Load Cod4.st. This represents a deterministic simulation in which a proportion, 0.2, of the stage 3 individuals is taken each year (see the Proportion of Individuals field on the Management & Migration sheet under the Model menu). There are 2,000 stage 3 individuals in the current population, so in the first year of operation, this harvest is the same as the fixed harvest applied in the file Cod2. Note that this harvest level is sustainable in the absence of environmental variation.

Step 5. Load Cod5.st. This represents the model in Cod4, with environmental and demographic variation added back in. Note that the total harvest is close to 12,500 kg, about the same total harvest as in the case in which a fixed number of individuals was harvested (in Cod3). The most striking difference between these results and the strategy involving a fixed harvest amount is that the mean population does not decline (see Trajectory Summary) and there is a negligible risk of the loss of the fishery (see Extinction/Decline).

Step 6. Repeat this simulation, using harvest proportions of between 0.1 and 0.6 of the stage 3 individuals per year. This can be done by modifying the Management & Migration sheet under the Model menu. Plot extinction risk versus the proportion of individuals harvested. Record the total harvest from each simulation.

Step 7. Plot extinction risk versus total harvest for both the constant harvest and the proportional harvest. Plot both curves on the same graph. What do these curves tell you about the effectiveness of these two strategies? Explain why many of the worlds fisheries are still managed using a fixed harvest amount. What problems might be introduced if the current population size was not known exactly, as it is in these simulations?

Exercise 8.3: Sustainable Use

Your goal in this exercise is to implement a harvesting strategy for the palm population that maximizes the dollar value of the resource over 50 years, provides for continuation of the resource 50 years, and provides a reasonable level of security of sustainable use, defined as the maintenance of a population of at least 1,000 individuals for the entire period.

Step 1. Develop a stage-structured model of the palm population. Use the stage matrix given above in Table 8.2. Specify values for the standard deviations the same as the means for the three fecundity values, in other words, a coefficient of variation of 100%. The upper-left element of the matrix (0.794) is not a fecundity, but the probability that a plant of smallest size class will remain in that size class the following year. Thus it is a survival rate, even though it is in the first row. For this element, and all other survival rates (numbers in other rows), specify the standard deviation as 10% of the mean.

Model the population's growth as exponential growth to a ceiling of 30,000 plants. Thus, in the density dependence screen, specify ceiling-type density dependence with the carrying capacity parameter equal to 30,000. The initial number of plants in each stage is assumed to be as follows:

    Stage 1: 3,000
    Stage 2: 1,000
    Stage 3:   300
    Stage 4:   300
    Stage 5:   300
    Stage 6:   100

Do not ignore the constraints, and use demographic stochasticity. After entering all the parameters, save the model. Run a stochastic simulation for 50 years, with 1,000 replications. When the simulation is finished, save the model again, this time with results.

The "Explosion/Increase" curve reports the chances of crossing an upper threshold at least once in the next 50 years. Thus, it records the chance that you will sell before 50 years are up, for different threshold values. Suppose your strategy is to sell all the palms if the population size exceeds 10,000 any time in the next 50 years, producing a two-fold return on the investment. What is the probability of this happening? What is the probability of reaching your target, if your target was a three-fold return, or a four-fold return?

Step 2. Remember that it was a requirement of the scenario that you manage the population sustainably. Definitions of sustainability vary. In this case, we shall define it as the requirement that you maintain a population of no less than 1,000 plants. If the population falls below that level, the land will default to the government and you will lose your asset. Thus the lower limit at 1,000 plants becomes an unacceptable lower bound, and judgment of alternative strategies must include an evaluation of the likelihood that you will cross it in the forthcoming 50 years.

What is the risk that the population size will fall below 1,000 plants at least once during the next 50 years?

Considering the probability of a two- or three-fold return, and the probability of failing to maintain 1,000 plants, do you consider this a wise investment?

Step 3. Suppose that at the beginning you have a further $10,000 in the bank. You may decide to use this capital to modify the processes to reduce the risk of failure, or to increase the expected size of the population. One of the options you have is to invest in seedling stock, artificially increasing the size of the population at the outset. The cost of obtaining, planting and caring for a seedling is $1; thus you can increase the initial number of individuals in stage 1 by 10,000.

Run another simulation that implements this option. Save the model and results in a new file. How does this change the answers to the questions in steps 2 and 3? Remember that this time you have $10,000 less in your bank account. This means that a four-fold return corresponds to 4 · ($50,000 + $10,000), or $240,000, which means selling the palms once the population is over 24,000 plants. For this question, assume that your target is a three-fold, or a four-fold return. A two-fold return (which requires selling 12,000 plants) does not make sense, since we start with 15,000 plants (5,000 that were already there, plus 10,000 additional seedlings).

How do the probabilities of three- or four-fold return, and the risk of failing to maintain 1,000 plants, change? Does this option change your mind about how wise this investment is?

Step 4. Another way you can spend your $10,000 is to set up an irrigation system to water the plants whenever the weather is dry. By doing so, you do not effect the initial population size, but you increase the mean value of the fecundities (because watering reduces the chances of mortality of seedlings during their first year), and also reduce their variability (because there will be less variation due to extreme drought years).

Load the first file you saved (without the additional 10,000 seedlings). Increase the three fecundities values by doubling their values, and decrease the standard deviation of each fecundity to 20% of its mean.

Run another simulation that implements this option. Save the model and results in a new file. How does this change the answers to the questions in steps 2 and 3? How do the probabilities of a two-, three- and four-fold return, and the risk of failing to maintain 1,000 plants, change? Does this option change your mind about the investment?

8.6 Further reading

Finkel, A. M. 1994. Stepping out of your own shadow: a didactic example of how facing uncertainty can improve decision-making. *Risk Analysis* 14:751–761.

Hilborn, R. 1987. Living with uncertainty in resource management. *North American Journal of Fisheries Management* 7:1–5.

Holling, C. S. 1993. Investing in research for sustainability. *Ecological Applications* 3:552–555.

Ludwig, D., Hilborn, R. and Walters, C. 1993. Uncertainty, resource exploitation, and conservation: lessons from history. *Science* 260:36.

Appendix:
RAMAS EcoLab Installation and Use

Requirements

The program requires an IBM-compatible computer running Windows 95, Windows 98, Windows NT 4.0, or later. The program will not work under Windows 3 or 3.1.

Memory: The computer should have at least 16 megabytes of memory. More memory would improve performance.

Processor: The program will run on an 80486 processor, although we recommend a Pentium or faster processor.

Hard disk space: The program requires approximately 2 megabytes of hard disk space.

Installation

If you received the program on a CD-ROM disc:
You must install the program on the hard disk; you cannot use RAMAS EcoLab from the disc. Put the CD-ROM disc in the CD-ROM drive. The installation program will start running. If it does not, select "Run" from the Start menu; type
    d:\setup.exe
where "d" is the letter of the CD-ROM drive, and press Enter. Follow the instructions on the screen.

If you received the program on floppy diskette(s):
Put the floppy diskette (#1, if there are more than one) in the floppy disk drive. Select "Run" from the Start menu; type
    a:\setup.exe
where "a" is the letter of the floppy disk drive, and press Enter. Follow the instructions on the screen.

    Store the distribution CD or diskette(s) in a safe place in case any of the program files are accidentally deleted.

    RAMAS EcoLab will be installed under your computer's "Program Files" folder. Double-click on the RAMAS EcoLab icon on your desktop to start the program. Press [F1] for help.

    You can also start RAMAS EcoLab from the "RAMAS EcoLab" group under "Programs" in the Start menu, or by double-clicking on the icons of associated data files (.SP, .ST, and .MP).

You might want to uninstall RAMAS Ecolab when you change computers or upgrade to a newer version of the program. You can do this by selecting "Uninstall" from the RAMAS EcoLab group under "Programs" in the Start menu. Note that this will delete all files that came with the program (including sample files). If you have made changes to any sample files that you'd like to keep, first copy them to a folder other than the folder where you initially installed RAMAS EcoLab (usually C:\Program Files\EcoLab).

Note: Read the file README.TXT for last-minute hints and corrections.

Using the program

See above for installing the program. Double-click on the RAMAS Ecolab icon on your desktop to start a shell program that provides access to all programs of RAMAS Ecolab. One of these, "Random numbers," lets you sample uniform random numbers for an exercise in Chapter 2. It gives a pair of uniform random numbers every time you click a button. The other choices are programs that let you build models:

> "Population growth" lets you build single population models with no age or stage structure (i.e., unstructured, or scalar models). These models can have variability (Chapter 2) or density dependence (Chapter 3).

> "Age and stage structure" lets you build single population models with age or stage structure, such as Leslie matrix models (Chapter 4) and stage matrix models (Chapter 5). These models can have variability and density dependence, as well as harvesting.

> "Multiple populations" lets you build metapopulation models with spatial structure (Chapter 6). These models can have variability, density dependence, and migration among populations.

The use of these three programs programs is very similar. Each program's main window consists of (1) title bar, (2) menu bar, (3) tool bar, (4) model summary, and (5) status bar.

(1) Title bar: At the top of the window is the title bar with the program name. On the title bar, at the upper-right corner of the window, are three buttons for minimizing, maximizing (or restoring to original size), and closing the main program window. Clicking the close button will terminate the program.

(2) Menu bar: Below the title bar is the menu bar, which includes six menus:

```
File  View  Model  Simulation  Results  Help
```

Click on one of these six words to open the pull-down menu. Alternatively, you can press the Alt key in combination with the underlined letter in the menu name. For example, pressing Alt-M will open the Model menu.

File menu is used to open or save model files. View menu is used to set display options. Selecting each item in the Model menu opens a dialog box that includes a group of model parameters. Simulation menu is used to run a simulation. After running a simulation, selecting each item in the Results menu displays one type of model result. The entries listed under Model and Results menus depend on the program. In each program, click "Help" to learn more about the operation of the program.

(3) Toolbar: Below the menu bar is the toolbar, which includes four buttons that can be used as shortcuts to access the following functions found under the File menu:

New (start a new model; same as pressing Ctrl-N)
Open (open an existing model; same as pressing Ctrl-O)
Save (save the model in a file; same as pressing Ctrl-S)
Exit (close the program; same as pressing Alt-X)

(4) Model summary: The largest part of the main program window contains a summary of the model. Depending on the program, this summary can take two forms:

> text, including title and comments (from the General information dialog), the number of replications, time steps, stages, and populations.
>
> map of the metapopulation.

(5) Status bar: At the bottom of the main program window is the status bar, which displays information about what the program is doing, as well as hints.

You can resize the program window by clicking on the lower-right corner of the window and dragging.

Some of the selections in the menus of a program (for example "Run") are procedures, and selecting them will make the program start computing. Others are *dialog boxes* for entering input parameters or displaying results. When you select one of the dialog boxes for input, the program will display a template on which you can type the values of the various *parameters*. After you enter your parameters, click "OK." If you want to leave a dialog box without making any changes to the input data, click "Cancel." The changes you have made since you opened the dialog box will be ignored. For help about input parameters, click "Help" (or press F1). The use of these programs are explained and demonstrated in the exercise sections of Chapters 2 through 6 (look under "RAMAS EcoLab" in the Index). Below, we discuss their general features.

Loading input files

In each program, you can load sample files. To do this, select **Open** from the File menu (or, press Ctrl-O), type in the filename or select a file by clicking.

Saving models and results

In each program, you can save a model you have created or modified. To do this, select **Save as** (to save a model with a different name) or **Save** (to save with the same filename) from the File menu. If you have already run the model, the results will also be saved.

Entering data

Within input windows under the Model menu (such as **General information**), you can type in parameter values, as well as title and comments. In all subprograms, the number of time steps (duration) and the number of replications are entered in **General information**. Setting replications to 0 is a convenient way of making the program run a deterministic simulation, even if the standard deviation of the growth rate is greater than zero.

When the number of replications is specified as 0, the program assumes a deterministic simulation and ignores parameters related to stochasticity. These parameters include the standard deviation matrix for age- or stage-structured models, and the parameters that are dimmed (not available for editing) in other input windows.

After editing an input window, click "OK" to accept the changes. (Note: clicking "Cancel" will close an input window without the changes you have made in that window.)

Erasing all input data and all results

To erase all input parameters and all results of a model, simply start a new model. You can do this by selecting **New** from the File menu.

Using the help facility

The function key F1 provides access to a context-sensitive help facility. You can press F1 or click the "Help" button anytime to get help about a particular window. In the help facility, you can get an overview of the help file by clicking on the "Contents" tab. In the Contents, click on a topic and then click "Open."

Running a simulation

After you have loaded a file, or created a model, you can run a simulation by selecting **Run** from the Simulation menu (or by pressing Ctrl-R). When the simulation starts, the program will open a Simulation window.

There are several controls on the toolbar at the top of the Simulation window. The first two buttons on the left (right under the word "Simulation" in the title) allow you to choose the simulation display (what to display during a simulation). By the default, the program will display trajectories or the metapopulation map, depending on the program.

For unstructured and age- or stage-structured models (Chapters 1 through 5), the program will display the population trajectory simulated by each replication. For metapopulation models (Chapter 6), the program will display a map of the metapopulation and will update the map at every time step.

The display of trajectories or maps may slow down the program. To turn off the display, click the first button from left on the toolbar. This will display only text (title, comments, and other parameters) during a simulation. This allows the simulation to be completed faster.

For more information, click the help button (with a ?) on the toolbar of the Simulation window.

When a simulation is completed, you will see "Simulation complete" at the bottom of the window. Close the Simulation window (click on the x in the upper-right corner) to return to the the main window. Once you return to the main window, you cannot go back to the display of individual trajectories (unless you run the simulation again).

Viewing and printing results

To view or print the results of a simulation, select one of the entries under the Results menu. This will open a window and display a graph. On top of the window is a series of buttons that

show a plot (display the result graphically, which is the default)

show numbers (display the result as a numerical table)

open a window for changing the scale and titles of the graph

save the result as a disk file

print the result (plot or text) on the default Windows printer

- copy the result to the clipboard, for pasting into another application
- display help for the particular result

For more information, click the help button (or press [F1]) and then click on "Copying, saving and printing results."

When a graph is displayed, the axes may have the letters k, m, or b. These indicate the multiplication factors:

k : × 1,000
m : × 1,000,000
b : × 1,000,000,000

Thus 2.50k means 2500 and 0.2m means 200,000.

Exiting the program

To exit from one of the subprograms, select **Exit** from the File menu (*Important: Remember to save your results before you exit*).

Technical support

User support from Applied Biomathematics is limited to technical aspects of using the program. The RAMAS home page has a list of frequently asked questions. If you want to contact us, please indicate the program and model you are using, describe the question or difficulty in detail, and if possible, attach a copy of the input file you were working on.

homepage: http://www.ramas.com
e-mail: ecolab@ramas.com
address: 100 North Country Road, Setauket, NY 11733 USA.

References

Akçakaya, H. R. 1990. Bald Ibis *Geronticus eremita* population in Turkey: An evaluation of the captive breeding project for reintroduction. *Biological Conservation* 51:225–237.

Akçakaya, H. R. 1991. A method for simulating demographic stochasticity. *Ecological Modeling* 54:133–136.

Akçakaya, H. R. 1992. Population viability analysis and risk assessment. In D. R. McCullough and R. H. Barrett (Eds.) *Wildlife 2001: Populations* (pp. 148–157). Elsevier Applied Science, London.

Akçakaya, H. R. 1994. GIS enhances endangered species conservation efforts. *GIS WORLD* Vol. 7, November 1994, pp. 36–40.

Akçakaya, H. R. 1998. *RAMAS GIS: Linking Landscape Data with Population Viability Analysis* (ver 3.0). Applied Biomathematics, Setauket, NY.

Akçakaya, H. R. and J. L. Atwood. 1997. A habitat-based metapopulation model of the California gnatcatcher. *Conservation Biology* 11:422–434.

Akçakaya, H. R. and B. Baur. 1996. Effects of population subdivision and catastrophes on the persistence of a land snail metapopulation. *Oecologia* 105:475–483.

Akçakaya, H. R. and M. Burgman. 1995. PVA in theory and practice. [letter] *Conservation Biology* 9:705–707.

Akçakaya, H. R. and L. R. Ginzburg. 1991. Ecological risk analysis for single and multiple populations. In A. Seitz and V. Loeschcke (Eds.) *Species Conservation: A Population-Biological Approach* (pp. 73–87). Birkhaeuser Verlag, Basel.

Akçakaya, H. R. and M. G. Raphael. 1998. Assessing human impact despite uncertainty: Viability of the northern spotted owl metapopulation in the northwestern USA. *Biodiversity and Conservation* 7:875–894.

Akçakaya, H. R., M. A. McCarthy and J. Pearce. 1995. Linking landscape data with population viability analysis: Management options for the helmeted honeyeater. *Biological Conservation* 73:169–176.

Allee, W. C. 1931. *Animal Aggregations: A Study in General Sociology*. University of Chicago Press, Chicago.

Allee, W. C., A. E. Emerson, O. Park, T. Park and K. P. Schmidt. 1949. *Principles of Animal Ecology*. Saunders, Philadelphia.

Andrewartha, H. G. and L. C. Birch. 1954. *The Distribution and Abundance of Animals*. University of Chicago Press, Chicago.

Askins, R. A. 1995. Hostile landscapes and the decline of migratory songbirds. *Science* 267:1956–1957.

Atwood, J. L. 1993. California gnatcatchers and coastal sage scrub: The biological basis for endangered species listing. In J. E. Keeley (Ed.) *Interface between Ecology and Land Development in California* (pp. 149–169). Southern California Academy of Sciences, Los Angeles.

Baars, M. A. and Th. S. van Dijk. 1984. Population dynamics of two carabid beetles at a Dutch heathland. I. Subpopulation fluctuations in relation to weather and dispersal. *Journal of Animal Ecology* 53:375–388.

Bernstein, B. B. and J. Zalinski. 1983. An optimum sampling design and power tests for environmental biologists. *Journal of Environmental Management* 16:35–43.

Beverton, R. J. H. and S. J. Holt. 1957. On the dynamics of exploited fish populations. *(Great Britain) Ministry of Agriculture, Fisheries and Food. Fishery Investigations* (series 2) 19: 5–533.

Bierzychudek, P. 1982. The demography of jack-in-the-pulpit, a forest perennial that changes sex. *Ecological Monographs* 52:335–351.

Birkhead, T. R. 1977. The effect of habitat and density on breeding success in the common guillemot (*Uria aalge*). *Journal of Animal Ecology* 46:751–764.

Bleich, V. C., J. D. Wehausen and S. A. Holl. 1990. Desert-dwelling mountain sheep: Conservation implications of a naturally fragmented distribution. *Conservation Biology* 4:383–390.

Botsford, L. W., T. C. Wainwright., J. T. Smith, S. Mastrup and D. F. Lott. 1988. Population dynamics of California Quail related to meteorological conditions. *Journal of Wildlife Management* 52:469–477.

Boyce, M. S. 1992. Population viability analysis. *Annual Review of Ecology and Systematics* 23:481–506.

Boyden, S. and S. Dovers. 1992. Natural-resource consumption and its environmental impacts in the western world: Impacts of increasing per capita consumption. *Ambio* 21:63–69.

Burgman, M. A. and V. A. Gerard. 1989. A stage-structured, stochastic population model for the giant kelp, *Macrocytis pyrifera*. *Marine Biology* 105:15–23.

Burgman, M., H. R. Akçakaya and S. S. Loew. 1988. The use of extinction models in species conservation. *Biological Conservation* 43: 9–25.

Burgman, M. A., S. Ferson and H. R. Akçakaya. 1993. *Risk Assessment in Conservation Biology*. Chapman and Hall, London.

Burnham, K. P. and D. R. Anderson. 1992. Data-based selection of an appropriate biological model: the key to modern data analysis. In D. R. McCullough and R. H. Barrett (Eds.) *Wildlife 2001: Populations* (pp. 16–30). Elsevier Applied Science, London.

Burnham, K. P., D. R. Anderson and G. C. White. 1996. Meta-analysis of vital rates of the northern spotted owl. *Studies in Avian Biology* 17:92–101.

Caswell, H. 1989. *Matrix Population Models: Construction, Analysis, and Interpretation.* Sinauer Associates, Sunderland, Massachusetts.

Caughley, G. 1994. Directions in conservation biology. *Journal of Animal Ecology* 63:215–244.

Crowder, L. B., D. T. Crouse, S. S. Heppell and T. H. Martin. 1994. Predicting the impact of turtle excluder devices on loggerhead sea turtle populations. *Ecological Applications* 4:437–445.

Cohen, J. E. 1995. Population growth and earth's human carrying capacity. *Science* 269:341–346.

Dempster, J. P. 1983. The natural control of butterflies and moths. *Biological Reviews* 58:461–481.

den Boer, P. J. 1968. Spreading of risk and stabilization of animal numbers. *Acta Biotheoretica* 18, 165–194.

Diamond, J. M. 1987. Extant unless proven extinct? Or, extinct unless proven extant? *Conservation Biology* 1:77–79.

Dodd, C. K. and R. A. Siegel. 1991. Relocation, repatriation, and translocation of amphibians and reptiles: are they conservation strategies that work? *Herpetologica* 47:336–350.

Ehrlich, P. R. and A. H. Ehrlich. 1990. *The population explosion.* Simon and Schuster, New York.

Elton, C. S. 1958. *The ecology of invasions by animals and plants.* Methuen, London.

Ferson, S., R. Akçakaya, L. Ginzburg and M. Krause. 1991. *Use of RAMAS to Estimate Ecological Risk: Two Fish Species Case Studies.* Technical Report EN-7176. Electric Power Research Institute, Palo Alto, California.

Finkel, A. M. 1994. Stepping out of your own shadow: a didactic example of how facing uncertainty can improve decision-making. *Risk Analysis* 14:751–761.

Fisher, R. A. 1930. *The Genetical Theory of Natural Selection.* Clarendon Press, Oxford.

Gause, G. F. 1934. *The Struggle for Existence.* Williams & Wilkins, Baltimore.

Gilpin, M. E. 1987. Spatial structure and population vulnerability. In M. E. Soulé (Ed.) *Viable Populations for Conservation* (pp. 126–139). Cambridge University Press.

Gilpin, M. E. and M. E. Soulé. 1986. Minimum viable populations: processes of species extinciton. In M. E. Soulé, (Ed.) *Conservation Biology: The science of scarcity and diversity* (pp. 19–34). Sinauer Associates, Sunderland, Massachussets.

Ginzburg, L. R., S. Ferson and H. R. Akçakaya. 1990. Reconstructibility of density dependence and the conservative assessment of extinction risks. *Conservation Biology* 4:63–70.

Griffith, B., J. M. Scott, J. W. Carpenter and C. Reed. 1989. Translocation as a species conservation tool: status and strategy. *Science* 245:477–480.

Gulland, J. A. 1971. The effect of exploitation on the numbers of marine animals. In P.J. den Boer and G.R. Gradwell (Eds). Proceedings of the Advanced Study Institute (pp. 450–468), Dynamics of numbers in populations, Oosterbeek, the Netherlands. 7–18 September 1970. Centre for Agricultural Publishing and Documentation, Wageningen.

Gunn, A., C. Shank and B. McLean. 1991. The history, status and management of muskoxen on Banks Island. *Arctic* 44:188–195.

Haefner, P. A. Jr. 1970. The effect of low dissolved oxygen concentrations on temperature-salinity tolerance of the Sand Shrimp, *Crangon septemspinosa* Say. *Physiological Zoology* 43:30–37.

Hanski, I. 1989. Metapopulation dynamics: does it help to have more of the same? *Trends in Ecology and Evolution* 4:113–114.

Hardin, G. 1993. *Living within limits: Ecology, economics and population taboos.* Oxford University Press, New York.

Harrison, S. 1991. Local extinction in a metapopulation context: an empirical evaluation. *Biological Journal of the Linnean Society* 42:73–88.

Hassell, M. P. 1986. Detecting density dependence. *Trends in Ecology and Evolution* 1: 90–93.

Hassell, M. P., J. Latto and R. M. May. 1989. Seeing the wood for the trees: detecting density dependence from existing life table studies. *Journal of Animal Ecology* 58:883–892.

Hilborn, R. 1987. Living with uncertainty in resource management. *North American Journal of Fisheries Management* 7:1–5.

Hilborn, R. and D. Ludwig. 1993. The limits of applied ecological research. *Ecological Applications* 3:550–552.

Hilborn, R. and C. J. Walters. 1992. *Quantitative fisheries stock assessment.* Chapman and Hall, New York.

Holdren, J. P. 1991. Population and the energy problem. *Population and Environment* 12:231–255.

Holling, C. S. 1993. Investing in research for sustainability. *Ecological Applications* 3:552–555.

Huenneke, L. F. and P. L. Marks. 1987. Stem dynamics of the shrub *Alnus incana* ssp. *rugosa*: transition matrix models. *Ecology* 68:1234–1242.

Hutchinson, G. E. 1957. Concluding remarks. *Cold Spring Harbor Symposium on Quantitative Biology* 22:415–427.

IGAE. 1992. The Inter-Governmental Agreement on the Environment (1992, parag. 3.5.1). Federal Government of Australia, Canberra.

IUCN 1994. Draft IUCN Red List categories. IUCN, Gland, Switzerland.

Jenkins, M. 1992. Species extinction. Pages 192–233 in *Global diversity: status of the earth's living resources.* World Conservation Monitoring Centre. Chapman and Hall, London.

Jenkins, S. H. 1988. Use and abuse of demographic models of population growth. *Bulletin of the Ecological Society of America* 69:201–202.

Keddy, P. A. 1981. Experimental demography of the sand-dune annual, *Cakile edentula*, growing along an environmental gradient in Nova Scotia. *Journal of Ecology* 69:615–630.

Keough, M. J. and G. P. Quinn. 1991. Causality and the choice of measurements for detecting human impacts on marine environments. *Australian Journal of Marine and Freshwater Research* 42:539–554.

Keyfitz, N. and W. Flieger. 1990. *World population growth and aging: demographic trends in the late twentieth century.* University of Chicago Press, Chicago.

Lack, D. 1966. *Population studies of birds.* Clarendon Press, Oxford.

LaHaye, W. S., R. J. Gutiérrez and H. R. Akçakaya. 1994. Spotted owl metapopulation dynamics in southern California. *Journal of Animal Ecology* 63:775–785.

Lamont, B. B., P. G. L. Klinkhamer and E. T. F. Witkowski. 1993. Population fragmentation may reduce fertility to zero in *Banksia goodii* — a demonstration of the Allee effect. *Oecologia* 94:446–450.

Lebreton, J.-D., R. Pradel and J. Clobert. 1993. The statistical analysis of survival in animal populations. *Trends in Ecology and Evolution* 8:91–95.

Lefkovitch, L. P. 1965. The study of population growth in organisms grouped by stages. *Biometrics* 21:1–18.

Leslie, P. H. 1945. On the use of matrices in certain population mathematics. *Biometrika* 33: 183–212.

Leslie, P. H. 1948. Some further notes on the use of matrices in certain population mathematics. *Biometrika* 35: 213–245.

Levins, R. 1970. Extinction. In M. Gerstenhaber (Ed.) *Some mathematical questions in biology.* American Mathematical Society, Providence, R.I.

Levinton, J. S. and L. Ginzburg. 1984. Repeatablity of taxon longevity in successive foraminifera radiations and a theory of random appearance and extinction. *Proceedings of the National Academy of Sciences* 81:5478–5481.

Ludwig, D. 1993. Environmental sustainability: magic, science, and religion in natural resource management. *Ecological Applications* 3:555–558.

Ludwig, D., R. Hilborn and C. Walters. 1993. Uncertainty, resource exploitation, and conservation: lessons from history. *Science* 260:17, 36.

Lyell, C. 1832. Principles of Geology, Vol. 2. Murray, London.

MacArthur, R. H. and E. O. Wilson. 1967. *The theory of island biogeography.* Princeton University Press, Princeton, N. J.

Mace, G. M. and R. Lande. 1991. Assessing extinction threats: toward a reevaluation of IUCN threatened species categories. *Conservation Biology* 5:148–1157.

Mapstone, B. D. 1995. Scalable decision rules for environmental impact studies: effect size, Type I, and Type II errors. *Ecological Applications* 5:401–410.

Margules, C. R. and M. P. Austin 1985. Biological models for monitoring species decline: the construction and use of data bases. In J. H. Lawton and R. M. May (Eds.) *Extinction rates* (pp. 183–196). Oxford University Press, Oxford.

Masaki, Y. 1977. Japanese pelagic whaling and whale sighting in the Antarctic, 1975–76. In International Whaling Commission, 27th report of the Commission (pp. 148–155). Office of the Commission, Cambridge.

McCarthy, M. A., D. C. Franklin and M. A. Burgman 1994. The importance of demographic uncertainty: an example from the helmeted honeyeater. *Biological Conservation* 67:135–142.

McCoy, E. D. 1995. The costs of ignorance. *Conservation Biology* 9:473–474.

McFadden, J. T., G. R. Alexander and D. S. Shetter. 1967. Numerical changes and population regulation in brook trout *Salvelinus fontinalis*. *Journal of the Fisheries Research Board of Canada* 24: 1425–1459.

Menges, E. S. 1990. Population viability analysis for an endangered plant. *Conservation Biology* 4:52–62.

Menges, E. S. and S. C. Gawler. 1986. Four-year changes in population size of the endemic Furbish's louseworth: implications for endangerment and management. *Natural Areas Journal* 6:6–17.

Moloney, K. A. 1986. A generalized algorithm for determining category size. *Oecologia* 69:176–180.

Morgan, M.G. and M. Henrion. 1990. *Uncertainty: a guide to dealing with uncertainty in quantitative risk and policy analysis.* Cambridge University Press, Cambridge.

Pearce, J. L., M. A. Burgman and D. C. Franklin. 1994. Habitat selection by helmeted honeyeaters. *Wildlife Research* 21:53–63.

Persson, L. and P. Eklov. 1995. Prey refuges affecting interactions between piscivorous perch and juvenile perch and roach. *Ecology* 76:70–81.

Pimm, S. L. 1991. *The Balance of Nature? Ecological Issues in the Conservation of Species and Communities.* The University of Chicago Press, Chicago.

Pimm, S. L., H. L. Jones, and J. Diamond. 1988. On the risk of extinction. *American Naturalist* 132:757–785.

Pinard, M. 1993. Impacts of stem harvesting on populations of Iriartea deltoidea (Palmae) in an extractive reserve in Acre, Brazil. *Biotropica* 25:2–14.

Pokki, J. 1981. Distribution, demography and dispersal of the field vole *Microtus agrestis* (L.) in the Tvärminne archipelago, Finland. *Acta Zoologica Fennica* 164:1–48.

Pollock, K. H., J. D. Nichols, C. Brownie and J. E. Hines. 1990. Statistical inference for capture-recapture experiments. *Wildlife Monographs* 107.

Possingham, H. P., D. B. Lindenmayer and T. W. Norton. 1993. A framework for the improved management of threatened species based on population viability analysis. *Pacific Conservation Biology* 1:39–45.

Pulliam, H. R. and N. M. Haddad. 1994. Human population growth and the carrying capacity concept. *Bulletin of the Ecological Society of America* 75:141–157.

Ricker, W. E. 1975. *Computation and Interpretation of Biological Statistics of Fish Populations*. Bulletin 191 of the Fisheries Research Board of Canada, Ottawa.

Robinson, S. K., F. R. Thompson III, T. M. Donovan, D. R. Whitehead and J. Faaborg. 1995. Reginal forest fragmentaion and the nesting success of migratory birds. *Science* 267: 1987–1990.

Rosenberg, A. A., M. J. Fogarty, M. P. Sissenwine, J. R. Beddington and J. G. Shepherd. 1993. Achieving sustainable use of renewable resources. *Science* 262:828–829.

Shaffer, M. L. 1981. Minimum population sizes for species conservation. *Bioscience* 31:131–134.

Shaffer, M. L. 1983. Determining minimum viable population sizes for the grizzly bear. *International Conference on Bear Research and Management* 5:133–139.

Shaffer, M. L. 1987. Minimum viable populations: coping with uncertainty. In M. E. Soulé (Ed.) *Viable populations for conservation* (pp. 69–86). Cambridge University Press, Cambridge.

Shaffer, M. L. 1990. Population viability analysis. *Conservation Biology* 4:39–40.

Simberloff, D., J. A. Farr, J. Cox and D. W. Mehlman. 1992. Movement corridors: conservation bargains or poor investments? *Conservation Biology* 6:493–504.

Smith, F. D. M., R. M. May, R. Pellew, T. H. Johnson, and K. S. Walker. 1993. Estimating extinction rates. *Nature* 364:494–496.

Sokal, R. R. and F. J. Rohlf. 1981. *Biometry*. Second ed. W.H. Freeman and Company, New York.

Solow, A. R. 1990. Testing for density dependence. A cautionary note. *Oecologia* 83:47–49.

Spencer, D. L. and C. J. Lensink. 1970. The muskox of Nunivak Island, Alaska. *Journal of Wildlife Management* 34:1–15.

Strong, D. 1986. Density-vague population change. *Trends in Ecology and Evolution* 1: 39–42.

Symonides, E., J. Silvertown and V. Andreasen. 1986. Population cycles caused by overcompansating density-dependence in an annual plant. *Oecologia* 71:156–158.

Thomas, C. D. 1991. Spatial and temporal variability in a butterfly population. *Oecologia* 87:577–580.

Usher, M. B. 1966. A matrix approach to the management of renewable resources, with special reference to selection forests. *Journal of Applied Ecology* 3:355–367.

Vandermeer, J. H. 1975. On the construction of the population projection matrix for a population grouped in unequal stages. *Biometrics* 31:239–242.

van Straalen N. M. 1985. Size-specific mortality patterns in two species of forest floor Collembola. *Oecologia* 67:220–223.

Verwijst, T. 1989. Self-thinning in even-aged natural stands of *Betula pubescens*. *Oikos* 56:264–268.

Vitousek, P. M. 1994. Beyond global warming: ecology and global change. *Ecology* 75:1861–1876.

Vitousek, P. M., P. R. Ehrlich, A. H. Ehrlich and P. M. Matson. 1986. Human appropriation of the products of photosynthesis. *BioScience* 36:368–373.

von Foerster, H., P. M. Mora and L. W. Amiot. 1960. Doomsday: Friday 13 November A. D. 2026. *Science* 132:1291.

Walters, C. J. 1985. Bias in the estimation of functional relationships from time series data. *Canadian Journal of Fisheries and Aquatic Sciences* 42:147–149.

WCMC. 1992. World Conservation Monitoring Centre. *Global diversity: Status of the earth's living resources.* Chapman and Hall, London.

Werner, P. A. and H. Caswell. 1977. Population growth rates and age *versus* stage-distribution models for teasel (*Dipsacus sylvestris* Huds.). *Ecology* 58: 1103–1111.

Woolfenden, G. E. and J. W. Fitzpatrick. 1984. *The Florida Scrub Jay: demography of a cooperative-breeding bird.* Princeton University Press, New Jersey.

Index

Acacia, 54
Age, 107, 138
  classes, 106
  units, 107
Age-structured model, 106
  assumptions, 107
  Brook Trout, 149
  composite age class, 107, 116
  demographic stochasticity, 123
  density dependence, 167
  environmental stochasticity, 125
  fecundities, 111, 113
  finite rate of increase, 121
  fishery management, 152
  human population, 148
  Helmeted Honeyeater, 108, 115, 144
  in RAMAS EcoLab, 144
  Leslie matrix, 113
  projections, 117
  reproductive value, 121
  sex ratio, 113
  stable age distribution, 119
  survival rates, 109
Allee effects, 86, 93
Animal species
  Blue Whale, 24, 25, 27
  Bluegill Sunfish, 80
  Brook Trout, 149
  Brown-headed Cowbird, 186
  California Gnatcatcher, 195, 226, 227, 234
  California Quail, 46
  Common Guillemot, 87
  Cotton-tail Rabbit, 21
  Florida Scrub Jay, 47
  Great Tit, 73
  Helmeted Honeyeater, 108, 115, 144, 191
  Human, 16, 28, 30, 148
  Japanese Beetle, 22
  Loggerhead Sea Turtle, 176
  Mountain Sheep, 184
  Muskox, 7, 62
  Northern Spotted Owl, 178
  Orchesella cincta, 130
  Perch, 4
  Sand Shrimp, 3
  Shrew, 56
  Silver-studded Butterfly, 192
  Spotted Owl, 94, 193, 210
  Starling, 21
Annual species, 8
Assumptions
  age-structured models, 107
  carrying capacity variation, 94
  correlated vital rates, 127
  density-dependent models, 91
  exponential model, 14
  life tables, 132
  metapopulation models, 198
  stage-structured models, 158

Baby boom, 119
Banksia goodii, 86
Benign demographic transition, 19
Beverton-Holt function, 84
billion, 16
Binomial distribution, 124
Births, 8
Blue Whale, 24, 25, 27
Bluegill Sunfish, 80
Brook Trout, 149

California Gnatcatcher, 195, 226, 227, 234
California Quail, 46
Capture-recapture, 134
Carrying capacity, 78, 89
  environmental variation, 94
  human population, 90
Catastrophes, 53
Ceiling model, 85
Census, 7, 108, 155
  post-breeding, 139
  pre-breeding, 138
Chaos, 91, 101
Climate change, 220
Cohort life table, 127, 132
Colonies, 87
Common Guillemot, 87
Composite age class, 107, 116, 159

Conservation, 35, 176, 199, 201, 209, 214, 222, 228
Constraints
  for Leslie matrix, 143
  for the stage matrix, 166
  on vital rates, 166
Contest competition, 76, 84
Correlation
  environmental fluctuations, 191, 201
  interaction with dispersal, 197
  vital rates, 127
Corridors, 201
Cost-benefit analysis, 228, 231
Cotton-tail rabbit, 21
Cowbird, 186
Critical, 223
Critically endangered, 222
Crowding, 72
Crucifer, 91
Cycles, 91

Deaths, 8
Demographic stochasticity, 38, 39, 44, 61
  fecundities, 125
  in age-structured models, 123
  in density-dependent models, 99
Density-dependent factors, 76
Density dependence
  affecting mortality, 72
  affecting reproduction, 73
  age-structured models, 167
  Allee effects, 86, 93
  Beverton-Holt function, 84
  ceiling model, 85
  contest, 76, 84
  demographic stochasticity, 99
  environmental variation, 94
  harvesting, 92, 102
  inverse, 86, 93
  logistic equation, 95
  parameter estimation, 97
  recruitment curve, 77
  replacement curve, 78, 84, 85
  Ricker equation, 80, 95
  scramble, 76, 78
  self-thinning, 74
  stage-structured models, 167

territories, 75
Deterministic simulation, 50
Diseases, 90
Dispersal, 193
  age-specific, 196
  corridors, 201
  density-dependent, 195
  direction, 194
  distance-dependent, 194
  interaction with correlation, 197
  stage-specific, 196
  stepping stone, 196
  stochasticity, 196
Doomsday prediction, 18
Doubling time, 12

Edge effects, 186, 187, 203
Eigenvalue, 121
Elasticities, 168, 181
Emigration, 8, 14
Endangered, 222, 223
Environmental variation, 45
  catastrophes, 53
  in age-structured models, 125
  in density-dependent models, 94
Erosion, 86
Error
  Type I, 243, 244
  Type II, 243
Exit, 272
Exotic species, 22, 220
Explosion, 51
Exponential decline, 23
  Blue Whale, 25
Exponential growth, 8
  applications, 16
  assumptions, 14
  doubling time, 12
  human population, 16
  long-lived species, 9
  pest populations, 21
Extinction, 214
  animal, 217, 218
  causes, 220
  Cretaceous, 215
  current rates, 216
  on islands, 216
  Permian, 215
  plant, 218

F1 (help), 270
Fecundity, 111, 113, 130, 138, 166
  annual species, 8
  estimating, 111, 135
Fertility, 130, 137, 140
Finite rate of increase, 8, 121, 131, 166
  estimating, 10
Fishery management, 152
Florida Scrub Jay, 47
Fluctuations
  age structure, 119
  density dependence, 91
  environmental, 45, 125
Fragmentation
  see Habitat fragmentation
Furbish's Lousewort, 185

Generation time, 131
Genetic engineering, 22
Geometric mean, 11
Global climate change, 220

Habitat fragmentation, 186, 202, 220
Habitat loss, 186, 220, 233, 234
Habitat management, 229
Harvesting, 14, 248
  decision rules, 253
  density dependence, 92, 102
  overexploitation, 220, 221
  reproductive value, 122
  simulation of, 153
Helmeted Honeyeater, 108, 115, 144, 191
Help facility, 270
Human population, 16, 28, 30
  age structure, 119, 148
  carrying capacity, 90
  energy use, 19, 21
Hypothesis testing, 242

Immigration, 8, 14
Impact
  detecting, 242
Inbreeding, 37, 86
Increase
  finite rate of, 8, 121, 131, 166
  instantaneous rate of, 131
Instantaneous rate of growth, 131

Interval extinction risk, 35
Island biogeography, 187
IUCN, 222

Jack-in-the-pulpit, 164
Japanese Beetle, 22

Kelp, 185

Lambda, 121
Leslie matrix, 113, 115, 140, 141
  Helmeted Honeyeater, 115
  projections with, 117
  reproductive value, 121
  stable age distribution, 119
Life tables, 127, 132
Local extinctions, 190
Loggerhead Sea Turtle, 176
Logging, 232
Logistic equation, 95
Long-lived species, 9

Malthus, 20
Mark-recapture, 134
Mass extinction, 215
Maternity, 130, 137, 140
Matrix multiplication, 114, 118, 163
Maximum rate of growth, 82
Measurement error, 36
Metapopulation, 184
Migration, 193
Minimum viable population, 213
Models, 5, 233, 258
  exponential, 9
Monitoring, 231
Mountain Sheep, 184
Multiple regression, 135
Muskox, 7, 62

Natural resource management, 241, 254
Net reproductive rate, 131
Niche, 2
Null hypothesis, 242

Orchesella cincta, 130
Overharvesting, 220, 221

Paramecium, 82, 83, 98

Parameter estimation
  density dependence, 97
  fecundity, 111, 113, 135, 136, 139, 140, 151
  Leslie matrix, 140, 141
  stage matrix, 173
  survival, 133, 134
  survival rate, 110, 117
  variance, 141, 142
  weighted averages, 133
Perch, 4
Plant species
  Acacia, 54
  Banksia goodii, 86
  Cakile edentula, 72
  Crucifer, 91
  Furbish's Lousewort, 185
  Jack-in-the-pulpit, 164
  Kelp, 185
  Palm, 248, 252
  Speckled Alder, 161
  Teasel, 174
  White Birch, 75
Poisson distribution
  fecundities, 125
Pollution, 220
Population, 2
  trajectory, 49
Population ceiling, 85
Population growth
  continuous time, 26
  doubling time, 12
  exponential, 8
  finite rate of increase, 8, 10, 121, 131, 166
  instantaneous rate of increase, 131
Population viability analysis, 213, 221
  components, 224
  limits, 232
Power, 243, 244
Precautionary principle, 247
Predator saturation, 86

Quasi-extinction, 44, 51
Quit, 272

Rainfall, 46, 47

RAMAS EcoLab, 62, 267
  age structure, 143
  Allee effects, 88
  constant harvest, 154
  constraints, 166
  correlation, 205
  deleting input data, 270
  density dependence, 98, 168
  deterministic simulation, 63, 144
  dispersal, 204
  entering data, 270
  erasing input and results, 270
  exit, 272
  extinction/decline, 64
  final age distribution, 146
  harvest, 153
  help facility, 270
  installation, 267
  loading files, 270
  map, 204
  metapopulation models, 203
  probabilities, 67
  proportional harvest, 153
  random numbers, 268
  running simulations, 271
  saving files, 270
  sensitivity analysis, 67
  standard deviation matrix, 147
  stochastic dispersal, 196
  stochastic simulation, 64
  technical support, 272
  trajectory summary, 63
  using, 268
  view/print results, 271
Random numbers, 268
Rate of growth
  maximum, 82
Rate of increase, 8, 121, 131, 166
Recolonization, 193
Recruitment, 80
Recruitment curve, 77, 80
Regression, 135
Reintroduction, 200
  reproductive value, 122
Replacement curve, 77, 78, 84, 85
Replication, 50, 59, 63, 144
Reproduction, 166
Reproductive value, 121
Reserve design, 201

Ricker equation, 80, 95
Risk, 35
Risk assessment, 35, 228
Risk curve, 44, 51, 64

Sample files
  California Gnatcatcher, 234
  California Spotted Owl, 193, 210
  human population, 148
Sand Shrimp, 3
Scramble competition, 76, 78
Sea Turtle, 176
Seed survival, 72
Self-thinning, 74, 253
Sensitivity analysis, 56, 67, 168, 178, 228
  deterministic, 168, 181
  management options, 170
  planning research, 168
  whole-model, 171
Sex ratio, 38, 113
Shrew, 56
Significance, 243, 244
Silver-studded Butterfly, 192
SLOSS, 201
Spatial heterogeneity, 53, 185
Species extinction risk, 190
Speckled Alder, 161
Spotted Owl, 94, 178, 193, 210
Spreadsheet, 151
Stable age distribution, 119, 121
Stage-structured model, 158
  assumptions, 158
  based on size, 159
  constraints, 166
  density dependence, 167
  diagram, 160
  finite rate of increase, 166
  Jack-in-the-pulpit, 164
  reproductive value, 166
  residence time, 165
  Sea Turtle, 176
  Speckled Alder, 161, 166

  stable distribution, 165
  Teasel, 174
Standard deviation, 39, 126, 150, 179
Standard error, 179
Starling, 21
Starting the program, 268
Static life table, 133
Stepping stone, 196
Stochastic simulation, 50
Stochasticity, 34
Survival rate, 109, 111, 117, 134, 166
  weighted averages, 133
Survivorship schedule, 128
Sustainability, 244, 256
Systematic pressure, 34, 221, 232

Teasel, 174
Technical support, 272
Terminal extinction risk, 35
Territoriality, 75, 77, 83
Threat categories, 222
Threshold, 35
Time to extinction, 59
Trajectory, 49
Transition rate, 164
Translocation, 13, 200

Uncertainty, 36, 241, 249, 256
  model, 55, 252
  parameter, 54
Uninstall, 268
Using RAMAS EcoLab, 268

Variability, 33, 37, 45, 49, 94
  see also Environmental variation
Variance, 39
  components, 141
  sums and products, 142
Vulnerable, 222, 223

Weighted average, 133
White Birch, 75
Worsening returns, 76